MECHANICAL ENGINEERING THEORY AND APPLICATIONS

INDUSTRIAL CONTROL SYSTEMS

MECHANICAL ENGINEERING THEORY AND APPLICATIONS

Additional books in this series can be found on Nova's website under the Series tab.

Additional E-books in this series can be found on Nova's website under the E-books tab.

ENGINEERING TOOLS, TECHNIQUES AND TABLES

Additional books in this series can be found on Nova's website under the Series tab.

Additional E-books in this series can be found on Nova's website under the E-books tab.

MECHANICAL ENGINEERING THEORY AND APPLICATIONS

INDUSTRIAL CONTROL SYSTEMS

ROBERT C. GILBERT
AND
ANGELA M. SCHULTZ
EDITORS

Nova Science Publishers, Inc.
New York

Library of Congress Cataloging-in-Publication Data

Industrial control systems / editors: Robert C. Gilbert and Angela M. Schultz.
p. cm.
Includes index.
ISBN 978-1-61209-988-0 (hardcover)
1. Adaptive control systems. I. Gilbert, Robert C. II. Schultz, Angela M.
TJ217.I53 2011
629.8--dc22

2011007085

Published by Nova Science Publishers, Inc. † New York

CONTENTS

PREFACE

In this new book, the authors present current research in industrial control systems. Topics discussed include the development of friction identification, modeling, and compensation methods for feed drive motions of CNC machine tools; MPC (Model Predictive Control) control algorithms for industrial applications; nondestructive dynamic monitoring of accelerated ion beams; Jacobian trajectory tracking for serial robot manipulators and intelligent control systems for an industrial manipulator.

Chapter 1 – It is a well-known fact that friction exists in feed drive servomechanisms and that it usually has some adverse effects. Such friction effects can be mitigated by model-based friction compensation methods, which are usually used in motion control design for CNC machine tools. Although there are many known experiments for identifying the friction characteristics of a feed drive servomechanism, all suffer significant limitations that affect the results, including position-dependent perturbations and unmodeled dynamics. In this chapter, a new friction extraction method was developed to reduce the effects of position-dependent perturbations, and a modified velocity control system was developed to reduce the effects of unmodeled dynamics. Moreover, simplified friction models that assume only constant static friction are usually used in motion control design in order to simplify implementation. However, the slideway normally used in feed drive servomechanisms of CNC machine tools usually induces significant amount of static friction, which is position-dependent and cannot be ignored for practical applications because it can significantly deteriorate the reversal motion of feed drive servomechanisms. The simplified friction model used for friction compensation limits the accuracy of feed drive motions in CNC machine tools. Therefore, an integrated friction model that considers both static friction and velocity-dependent friction was developed in this chapter for friction compensation in CNC machine tools. Several experiments and motion tests were carried out on a three-axis CNC milling machine to illustrate the feasibility of the proposed friction identification method. The experimental results indicate that the developed friction model achieves a 44.16% reduction in the root-mean-square value of tracking errors in comparison to an uncompensated motion control system Moreover, as compared to the conventional identification method, a further reduction of 9.52% is achieved in the root-mean-square value of tracking errors. Because the payload conditions that exist in the feed drive servomechanism of a CNC machine tool could significantly affect the friction characteristics, this chapter also analyzed the friction characteristics under different payload conditions that can be caused in a feed drive servomechanism. A feedforward friction torque compensation structure is usually used for friction compensation on the basis of the given position commands, velocity commands, and an integrated friction model. However, for mechanical systems that are actuated by velocity-controlled servomotor packs, the torque compensation structure cannot be applied directly to reduce the effects of friction, and thus a velocity-based friction compensator is required in the motion control design. Therefore, a new velocity-based friction compensation structure with an additional velocity input was developed in this chapter to overcome the adverse effects induced by friction to an extent. In the compensation structure, a prefilter is developed to generate an additional velocity command; moreover, the parameters of the prefilter can be fine-tuned to reduce the motion errors. A number of circle motion tests with different feedrates and loading conditions were carried out on a three-axis CNC milling machine in order to illustrate the feasibility of the proposed friction compensation method. The experimental results indicate that a maximum reduction of 58.71% is achieved in the absolute contouring errors during circular motions by the friction modeling and compensation methods developed in this chapter.

Chapter 2 – The MPC (Model Predictive Control) control algorithms have been well accepted for industrial applications. Among the advantages of these algorithms is the fact that they deal directly with interactions between inputs and outputs and are easily applied to a multivariable process (MIMO - Multiple Input Multiple Output). In addition, restrictions on inputs and outputs due to physical constraints and security, common in process industries, can be directly integrated into the MPC algorithm. One of the most popular predictive control algorithms is the DMC or dynamic matrix control, which is based on a linear model. One difficulty in using linear control algorithms is that the majority of the chemical processes exhibit a nonlinear behavior. Thus, controllers based on linear models must have their parameters adjusted in a conservative manner to ensure stability in a wide range of operating conditions, which can result in a very slow control. In some cases, even with this conservative adjustment of parameters, the linear controller does not provide satisfactory performance. Some examples where this can occur are non-stationary processes in which an optimal profile is implemented, systems that exhibit complex interactions with constraints and highly nonlinear phenomena, and processes subject to large disturbances. Another option for improving the performance of the control scheme is the use of predictive control algorithms based on nonlinear models. Several types of models have been used for this purpose, such as differential equations, neural networks, etc. Artificial neural networks have been increasingly applied to describe chemical processes due to their ability to approximate highly nonlinear systems involving a large number of state variables. Different structures of neural networks were considered as non-linear models for advanced control algorithms. In this chapter, the goal is to study the nonlinear control of a continuous extractive process of bioethanol production. An algorithm based on DMC is proposed, using an internal model based on artificial neural networks to represent the dynamic behavior of the manipulated and controlled variables, in substitution of the linear convolution model of the DMC controller. The extractive alcoholic fermentation process combines the fermentor and a flash vessel to continuously extract part of the ethanol produced. Experimental data from 400 h of operation of a pilot plant is used to develop the ANN based model for the control system.

Chapter 3 – A complex for dynamic nondestructive diagnostics of accelerated ion beams has been developed. The complex includes the detector, measurement, data collection systems and the software for beam control and optimization of irradiation regimes. The system of dynamic diagnostics of film irradiation based on transparent (95% transparency) profilometers for track membrane production was developed and implemented at the accelerator complex "Alpha" (JSC "Trackpore Technology", Moscow, Russia). This system provides measurement, dynamic control, and certification of irradiation parameters with a time resolution of 100 μs. Industrial track membrane production regimes with given irradiation density and homogeneity were studied. Practical recommendations on optimal regimes were formulated as part of the metrological control of industrial film irradiation technology. The system of dynamic monitoring of space-time characteristics of circulating beam based on microchannel plates was developed for the superconducting accelerator Nuclotron (Joint Institute for Nuclear Research, Dubna, Russia). The system possesses unique time characteristics, it provides measurement of nuclear beam intensities during acceleration in a range from 10^4 to 10^{11} pps.

Chapter 4 – To overcome the singularities and uncertainties in arm configurations resulting from applying the inverse kinematics model for serial robot manipulators, adaptive learning algorithm using an artificial neural network (ANN) was proposed in this chapter to solve the problem. The proposed technique does not require any prior knowledge of the kinematics model of the system being controlled, the main idea of this approach is the use of an ANN to learn the robot system characteristics rather than having to specify an explicit robot system model. In this approach a network was trained to learn desired set of angular positions and velocities from a given set of end effector's positions and velocities. Training was performed Off-line and training data were recorded from sensors fixed on each joint. Testing the resulting network on a general 6 DOF serial robot manipulator has shown the generality and efficiency of the proposed algorithm.

Chapter 5 – In this chapter, two types of intelligent control systems are introduced. One is the adaptive learning technique for large-scale teaching signals. Recently various control methods as represented by PID control are used for robotic control. To cope with the requirements for high response and precision, advanced feedforward controllers such as gravity compensator, Coriolis/centrifugal force compensator and friction compensators have been built in the controller. Generally, it causes heavy computational load when calculating the compensating value within a short sampling period. To overcome the problem, integrated recurrent neural networks are applied as a feedforward controller for PUMA560 manipulator. The feedforward controller works instead of gravity and Coriolis/centrifugal force compensators. In the learning process of neural network by using back propagation algorithm, in general, the learning coefficient and gain of sigmoid function are tuned intuitively and empirically according to teaching signals. The tuning is complicated because of being conducted by trial and error. Especially, when the scale of teaching signal is large, the problem becomes crucial. To cope with the problem concerning the learning performance, a simple and adaptive learning technique for large-scale teaching signals is proposed. The learning techniques and control effectiveness are evaluated through simulations using the dynamic model of PUMA560 manipulator. The other is the fine gain tuning technique for model-based robotic servo controllers using genetic algorithms. Resolved acceleration control method and computed torque control method are used for nonlinear control of industrial manipulators, which are composed of a model base portion and a servo portion. The servo portion is a closed loop system with respect to the position and velocity. On the other hand, the model base portion has the inertia term, gravity term and centrifugal/Coriolis term, which works for canceling the nonlinearity of manipulator. In order to realize high control stability, the position and velocity feedback gains used in the servo portion should be tuned suitably. In the latter half of the chapter, a simple but effective fine tuning method after the manual tuning process is introduced for the position and velocity feedback gains in the servo portion. At the first step, the search space for the gains are roughly narrowed down by a controller designer, e.g., considering the critically damped condition. At the next step, the gains are finely tuned within the space by using genetic algorithms. The genetic algorithms search for the better combination of the position and velocity gains. Simulations are conducted using the dynamic model of PUMA560 manipulator to validate the effectiveness of the proposed method.

Chapter 6 – In this work, Global Automata are presented, as a method offering the convenience of modeling various types of industrial systems without any restrictions on systems properties. Their structure has common characteristics with several types of automata, such as the control graph with a finite set of states and transitions between those states. Global automata can be used for modeling hybrid systems handling both discrete and real valued variables combining flow, invariant and guard conditions from hybrid automata, with clock constraints and delayed inputs from timed and PLC automata. In addition, new modeling parameters, as reset table at each transition and hierarchical classification of executable events at each state, are introduced. Application independence derives from the fact that they are a super-set of every other type of automata and hence are less application depended compared to any of them. Based on Global Automata, some tools were developed in order to increase their modeling power. First tool is the ability of state aggregation, which generally means the merging of two or more states in order to produce a new "super-state". State aggregation is a powerful tool to avoid state space expansion. The second developed tool is automata composition and is mainly used in cases where ad-hoc modeling is not a practical solution, since it allows the construction of large system model using simple models of all system's components. Finally, global automata allow the development of hierarchical models, consisting of two or more automata. In such case, the automaton of the higher level has the role of supervisor feeding the automata of lower level with data, reference values etc. Two implementation tools for programming industrial controllers are also presented. The first tool is a synthesis tool for translating a global automaton into executable code of a programming language (e.g. C, Matlab). Tool usage is important for simulation and verification but is not limited on this. Contemporary industrial systems are equipped with Supervisory Control and Data Acquisition Systems (SCADA), which have embedded functions of running code. Consequently, an industrial system modeled as a global automaton can be controlled by the executable code running in a SCADA station. However in industry the dominant controller is the Programmable Logic Controller, which is a special aim computer suitably built for the application control tasks. The IEC 61131-3 Programming Norm describes all the well-known languages for programming PLCs and the second tool is a synthesis tool for implementing global automata in PLCs based on this programming norm. Global automata, as a modeling method, and all the above mentioned tools are illustrated through representative examples. The presented examples are either theoretical or real systems implemented in the laboratory or even real production systems.

In: Industrial Control Systems
Editors: R. Gilbert and A. Schultz
ISBN: 978-1-61209-988-0

Chapter 1

Development of Friction Identification, Modeling, and Compensation Methods for Feed Drive Motions of CNC Machine Tools

Syh-Shiuh Yeh*

Department of Mechanical Engineering, National Taipei University of Technology, Taipei, Taiwan

Abstract

It is a well-known fact that friction exists in feed drive servomechanisms and that it usually has some adverse effects. Such friction effects can be mitigated by model-based friction compensation methods, which are usually used in motion control design for CNC machine tools. Although there are many known experiments for identifying the friction characteristics of a feed drive servomechanism, all suffer significant limitations that affect the results, including position-dependent perturbations and unmodeled dynamics. In this chapter, a new friction extraction method was developed to reduce the effects of position-dependent perturbations, and a modified velocity control system was developed to reduce the effects of unmodeled dynamics. Moreover, simplified friction models that assume only constant static friction are

* Corresponding author. Tel.: +886-2-2771-2171 ext. 2025; fax: +886-2-2731-7191. E-mail address: ssyeh@ntut.edu.tw (Syh-Shiuh Yeh).

usually used in motion control design in order to simplify implementation. However, the slideway normally used in feed drive servomechanisms of CNC machine tools usually induces significant amount of static friction, which is position-dependent and cannot be ignored for practical applications because it can significantly deteriorate the reversal motion of feed drive servomechanisms. The simplified friction model used for friction compensation limits the accuracy of feed drive motions in CNC machine tools. Therefore, an integrated friction model that considers both static friction and velocity-dependent friction was developed in this chapter for friction compensation in CNC machine tools. Several experiments and motion tests were carried out on a three-axis CNC milling machine to illustrate the feasibility of the proposed friction identification method. The experimental results indicate that the developed friction model achieves a 44.16% reduction in the root-mean-square value of tracking errors in comparison to an uncompensated motion control system Moreover, as compared to the conventional identification method, a further reduction of 9.52% is achieved in the root-mean-square value of tracking errors. Because the payload conditions that exist in the feed drive servomechanism of a CNC machine tool could significantly affect the friction characteristics, this chapter also analyzed the friction characteristics under different payload conditions that can be caused in a feed drive servomechanism. A feedforward friction torque compensation structure is usually used for friction compensation on the basis of the given position commands, velocity commands, and an integrated friction model. However, for mechanical systems that are actuated by velocity-controlled servomotor packs, the torque compensation structure cannot be applied directly to reduce the effects of friction, and thus a velocity-based friction compensator is required in the motion control design. Therefore, a new velocity-based friction compensation structure with an additional velocity input was developed in this chapter to overcome the adverse effects induced by friction to an extent. In the compensation structure, a prefilter is developed to generate an additional velocity command; moreover, the parameters of the prefilter can be fine-tuned to reduce the motion errors. A number of circle motion tests with different feedrates and loading conditions were carried out on a three-axis CNC milling machine in order to illustrate the feasibility of the proposed friction compensation method. The experimental results indicate that a maximum reduction of 58.71% is achieved in the absolute contouring errors during circular motions by the friction modeling and compensation methods developed in this chapter.

Keywords: Friction identification; Friction modeling; Friction compensation; Feed drive motions; CNC machine tools

1. INTRODUCTION

Friction is a physical phenomenon that exists in mechanical systems. It usually causes some adverse effects such as steady-state errors, stick-slip oscillations, and poor tracking and contouring performances in the motion control of these mechanical systems. To reduce these adverse effects and achieve high precision motions, model-based friction compensators are usually used in the design of motion control systems to compensate for the friction phenomenon [1-4]. Therefore, good friction models are usually required in motion control design, and the extent of friction-compensation heavily depends on the accuracy of the applied friction models.

A good friction model is usually based on the knowledge of friction behaviors in a mechanical system. However, it is difficult to model these friction behaviors because many factors affect them, including the material, lubrication, and surface texture of the bodies in contact [5-13]. Thus, numerous simplified friction models have been proposed for motion control and compensation design [3-5,12-13]. For a given friction model, an identification process, usually based on an optimization method, is then applied to identify the parameters of the model. For instance, Cheok et al. [14] used the simplex method to determine the parameters of the Karnopp friction model [15]. Kim et al. [16] proposed an acceleration evolutionary programming method, which is an experimental identification method, to identify the parameters of the seven-parameter friction model presented by Armstrong-Helouvry et al. [4]. For the given friction-velocity map obtained by experiments, Canudas-de-Wit and Lischinsky [17] also used the simplex method to obtain the optimal parameters for the LuGre friction model [18].

Conventionally, two experiments, the breakaway experiment and the constant-velocity experiment, are carried out to obtain the friction characteristics of a mechanical system [12,17,19]. The breakaway experiment consists of allowing the moving part of the mechanical system to come to rest at certain positions, ramping up the driving torque, and then measuring the torque required to initiate motion of the moving part. Therefore, the breakaway experiment, which samples static friction with a suitable spatial frequency, can be used to identify position-dependent friction and develop a friction compensation lookup table to compensate for the same. The constant-velocity experiment can be used to identify velocity-dependent friction of a mechanical system by mapping the friction and velocity values measured during constant-velocity motion of the moving part of the system. To reduce the adverse effects of velocity noise and ripple torque, the friction and velocity values used to create the friction-velocity map are obtained by averaging the measured velocity and input torque values. In this experiment, the actuator used to drive the moving part of the mechanical system is actuated by employing velocity feedback control using a velocity PI controller to obtain good velocity tracking results and to support the assumption that the velocity responses remained constant during steady motions. For a given friction-velocity map and friction model, linear or nonlinear optimization algorithms are then used to estimate the parameters of the friction model by fitting the data to the friction-velocity map [12,17].

Although the existing experiments can be used to obtain the friction characteristics of mechanical systems, some problems that could significantly affect the friction identification results, still exist. For instance, most of the friction models of mechanical systems take into account only two contact surfaces. However, in practice, several contact surfaces exist in mechanical systems, all of which could generate friction with different characteristics. Therefore, for mechanical systems constructed with many and different mechanical parts, it is difficult to model friction characteristics by a friction model with simplex characteristics. Furthermore, in order to construct the friction-velocity map of a mechanical system, its velocity responses must be kept constant during steady motions. However, because there are considerable position-dependent perturbations in a mechanical system, usually caused by the surface irregularities, assembly errors, and geometric machining errors in the mechanical parts, the velocity responses always fluctuate and cannot be kept constant for any length of time. These position-dependent perturbations could deteriorate the obtained friction-velocity map; the unmodeled dynamics, usually induced by several flexible and inertial elements in the mechanical system, increase the uncertainties in the estimated friction characteristics. Therefore, in order to obtain reasonable friction and velocity data in the presence of these unavoidable perturbations, it is necessary to modify the velocity feedback control system used in the constant-velocity experiment by means of robust control. The disturbance observer (DOB), introduced by Ohnishi [20], is usually applied to motion control systems to compensate for external disturbances and to reduce the adverse effects of system uncertainties and nonlinearities. Because the observed value include information integrated from different friction effects, the DOB has also been used to identify the friction in a case where the inertia of the applied system is obtained accurately and no other external disturbances exist in the system [21]. However, position-dependent perturbations always exist in mechanical systems, and these could affect the friction identification results of the proposed identification methods based on the DOB structure.

In this chapter, because of its merits [20], the DOB feedback control structure is integrated into a conventional control system that uses only a PI controller for velocity feedback control. Because the modified velocity feedback control system can compensate for any speed perturbations induced by external disturbances and system uncertainties, the velocity responses of the applied servomechanism can be kept constant during steady motions in the constant-velocity experiment. Moreover, because the position-dependent perturbations in a mechanical system, which are usually constant under invariant mechanical conditions, can be related to the absolute position of a moving part in the mechanical system, a new friction extraction method is developed to remove these position-dependent perturbations by considering the subtraction of different frictions at the same absolute positions of the moving part. The friction and velocity data values obtained using the modified velocity control system and the newly-developed friction extraction method can thus be used to develop a friction model that reduces the uncertainties in the estimated friction characteristics. In order to complete the friction model proposed in this chapter, the breakaway experiment [12] is employed to estimate static frictions, and the obtained results are then applied to develop friction compensation lookup tables. Several experiments and motion tests were carried out on a three-axis CNC milling machine to demonstrate the feasibility of the approaches proposed in this chapter.

A feedforward friction compensation structure is usually employed to control the velocity and position of the motion in a mechanical system. The torque value used for friction compensation is computed by referring to the given velocity commands and a friction model. However, in some cases, the friction torque compensation structure cannot be directly applied to control feed drive servomechanisms that are actuated by velocity-controlled servomotor packs. Therefore, a velocity-based friction compensator is needed to reduce the adverse motions induced by frictions. In this chapter, by referring to the existing feedforward friction torque compensation structure, a new velocity-based friction compensation structure is developed to reduce the friction effects by employing an additional velocity command. Here, the additional velocity command for friction compensation is obtained from a prefilter using input signals computed by referring to the position command, velocity command, and the friction model obtained in this chapter. In this chapter, the prefilter is designed as a low-pass-filtered derivative operator under the assumption that the velocity feedback controller is a PI controller.

However, for a velocity feedback controller with an unknown structure, the parameters of the prefilter can first be systematically identified and then fine-tuned manually to minimize the adverse effects caused by friction. Several circular motion tests were carried out on a three-axis CNC milling machine to show the feasibility of the compensation method proposed in this chapter.

This chapter is organized as follows. Section 2 describes the design of the experiment and measurement method used in this chapter to obtain data values for estimating the frictions of the applied feed drive servomechanisms. This section also details the experimental setup employed to verify the modeling method proposed in this chapter, the velocity control structure used to control the feed drive servomechanisms in the experimental setup, and the breakaway and constant-velocity experiments usually used to measure frictions. Section 3 presents the friction identification techniques proposed in this chapter and reports the results of this identification. This section reports the spectral analysis results of the obtained position-dependent frictions and also presents the data extraction method developed to obtain reliable friction and velocity data values. A new friction-velocity map is introduced in this section. This section also details the methods used to develop friction compensation lookup tables and velocity-dependent friction models to compensate for different frictions. Section 4 presents the design of the velocity-based friction compensator proposed in this chapter. The design and tuning method of the prefilter are also described in detail in this section. Section 5 shows the experimental results of sinusoidal motion tests to show the feasibility of the friction identification method proposed in this chapter. Section 6 shows the experimental results of circle motion tests under different loading conditions by applying circle motion commands with different feedrates. The experimental results show the feasibility of the friction compensation method proposed in this chapter. Section 7 concludes this chapter.

2. Descriptions of Experiment Design and Measurement

The aim of this section is to describe the method proposed in this chapter for identifying the position-dependent and velocity-dependent frictions that affect the motions of the feed drive servomechanism of a CNC machine tool. This section first presents the experimental setup used in this chapter and then presents the modified velocity feedback control system used in the experiments. Then, two experiments, the breakaway experiment and constant-velocity experiment, are introduced to obtain friction and velocity data values for the estimation of position-dependent and velocity-dependent frictions.

2.1. Experimental Setup

Experimental studies were performed on a three-axis CNC milling machine comprising an industrial PC, DSP-based motion control card, and mechanical system with commercial servomotor packs. Figures 2.1 and 2.2 show a schematic drawing and photograph of the experimental setup, respectively. An industrial PC with a Pentium IV 2.8 GHz CPU is used to provide various functions, including a human-machine operator interface, the implementation of identification methods, and the recording of data values during the experiments. A digital-signal-processor-based (DSP-based) motion control card with a high-performance TI TMS320F2812 digital signal processor (DSP) is used as the interface for sending motion commands to and for receiving feedback signals from the applied servomotor packs with a sampling period of 1.0 msec.

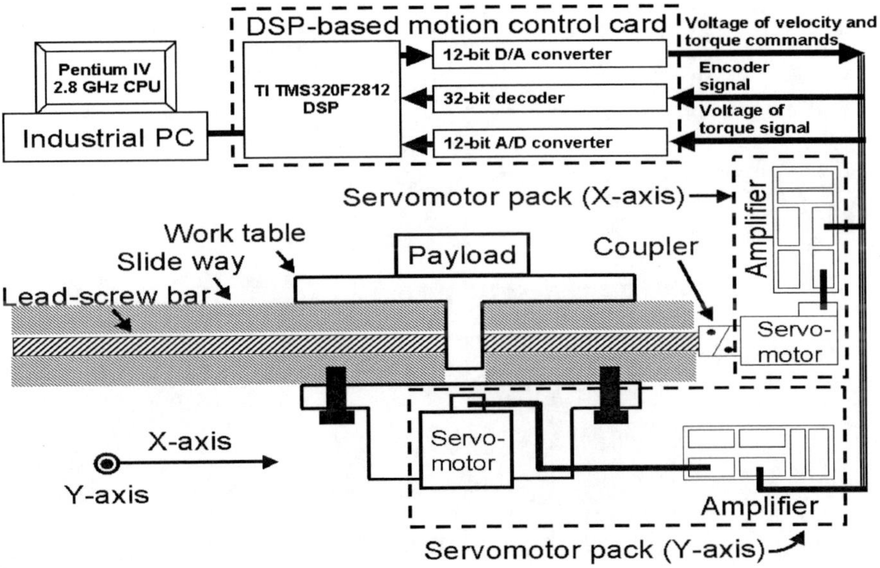

Figure 2.1. Schematic drawing of experimental setup.

The mechanical system of the applied milling machine is mainly composed of a vertical axis and biaxial moving table. The vertical axis with a rotary spindle can move a cutting tool vertically to perform machining with different depths. The biaxial moving table, which is the main object of this chapter, is driven by two PANASONIC A4-type AC servomotor packs equipped with built-in velocity and torque control loops. Here, torque and velocity command signals are sent to each amplifier of the servomotor packs through a 12-bit D/A converter on the DSP-based motion control card. A rotary incremental encoder is directly coupled to the servomotor of a servomotor pack, and the generated encoder signals, which are used to indicate the absolute angular positions of the servomotor, are received through a 32-bit decoder on the DSP-based motion control card. The angular velocity signal used in this chapter is computed by angular position differentiation over a sampling period. A 12-bit A/D converter implemented on the DSP-based motion control card is used to receive analog signals from servomotor packs to obtain the torque values that actuate the biaxial moving table. MATLAB software is used in this chapter to facilitate the development of identification and control algorithms.

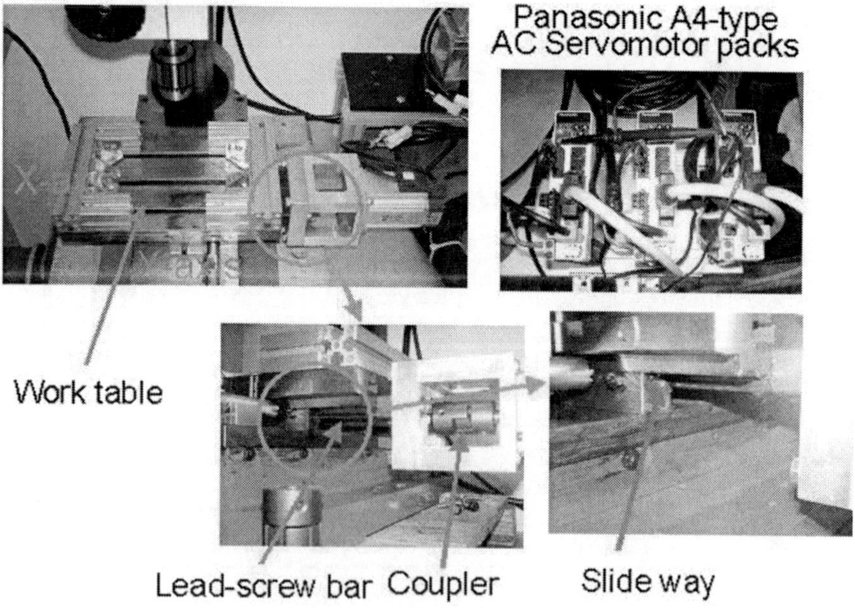

Figure 2.2. Photograph of experimental setup.

The biaxial moving table mainly consists of a moving part and two feed drive servomechanisms, as shown in Figures 2.1 and 2.2. The X-axis, along with a moving part, i.e., a work table, is installed on the Y-axis. The moving part is placed on a feed drive servomechanism to form the X-axis axial motion system of the CNC milling machine. Here, the moving part, which is placed on a slide way, is driven by a servomotor pack through a coupler, lead-screw bar, and nut. The moving part and the support rails of the slide way are both made of cast iron, and therefore lubrication must be guaranteed before performing experiments. The X-axis axial motion system is placed on another feed drive servomechanism to form the Y-axis axial motion system of the CNC milling machine. Taken together, the X-axis and Y-axis axial motion systems constitute the orthogonal and planar coordinate system for the motion of the CNC milling machine. The payload of the biaxial moving table can be modified by adding weights in increments of 1 kg. Consequently, several loading conditions, from an empty table to a table with a 5 kg payload, are obtained in this chapter. Table 2.1 shows some of the specifications of the main components used in this chapter.

Table 2.1. Specifications of main components in this chapter

Milling machine	Work table area (mm × mm)	240 × 145
	Size (mm × mm × mm)	500 × 350 × 690
	X/Y/Z stroke (mm × mm × mm)	130 × 180 × 230
	Maximum speed (mm/min)	1,200
	Screw lead (mm)	1.2
Servomotor pack	Rated power (kW)	0.2
	Rated torque (Nm)	0.64
	Maximum torque (Nm)	1.9
	Rated angular velocity (rpm)	3,000
	Maximum angular velocity (rpm)	5,000
	Inertia of rotor (kg-m^2)	0.42×10^{-4}
	Resolution of encoder (encoder increments/revolution)	10,000
	Resolution of encoder (radian/encoder increment)	6.2832×10^{-4}

2.2. Measurement and Control

The measurement of the friction and velocity data values was performed during motions of the moving part placed on the feed drive servomechanism of the CNC milling machine. Figure 2.3 schematically shows the motions of the moving part, i.e., the work table, during the experiments discussed in this chapter. The moving part starts moving forward from return point A, sequentially passes through the home position and target position, and reaches return point B. It then moves backward from return point B and reaches return point A. For each experiment, the moving part moves several times forward and backward along the moving path with different motion velocities. The moving paths between return position A and the home position and between return position B and the target position must be sufficiently long to ensure that the movement between the home position and target position is steady state and backlash free. Therefore, the identified position-dependent and velocity-dependent frictions are valid when the moving part moves along the path between the home position and target position. In this chapter, the absolute positions of the moving part are represented by the encoder increments obtained from the rotary encoder coupled to the servomotor of the applied servomotor pack. The home position is located at the zero encoder increment and the target position is located at 150,000 encoder increments. Several iterations of the cyclic motions between return position A and return position B are adopted in each experiment to obtain the average behaviors of the frictions. An infrared thermometer is employed to measure the temperature around the moving table, and this temperature is controlled at 28–29 °C during all of the experiments. The industrial PC records data values with a period of 10 msec, and the acquired data include the motion direction, absolute position, velocity, acceleration of the moving part, torque value that actuates the tested feed drive servomechanism, and motion time. The units of the acquired absolute position, velocity, and acceleration signals are encoder increments, rad/sec, and rad/sec2, respectively. The unit of the acquired torque value is Nm. Usually, a huge number of data points is required to obtain good identification results, and thus a long experiment period (normally several hundred hours) is mandatory.

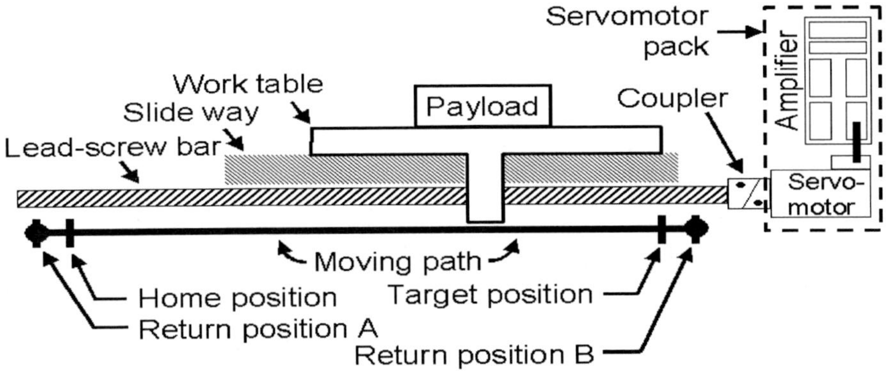

Figure 2.3. Motions of moving part (work table) placed on feed drive servomechanism.

Figure 2.4 shows the modified velocity feedback control structure used in this chapter to keep the velocity responses of the applied feed drive servomechanism constant during steady motions in the constant-velocity experiment. This control structure is composed of a built-in velocity feedback control loop and a disturbance observer. Each of the applied servomotor packs provides a built-in velocity control loop with a PI feedback controller. The PI controller generates actuating torque τ_m to overcome external torque τ_{ext} and then actuate a feed drive servomechanism with an equivalent inertia, J_a. In this chapter, the external torque, τ_{ext}, only consists of position-dependent and velocity-dependent frictions, without considering other external disturbances such as cutting forces. ω_{cmd} and ω denote the angular velocity command and angular velocity response of the applied servomotor, respectively. Here, the proportional gain and integral gain of the PI controller are tuned using the built-in auto-tuning method [22] to make the feed drive servomechanism obtain good velocity tracking results. A disturbance observer (DOB) is integrated into the built-in velocity control loop to reduce the adverse effects induced by the uncertainties and nonlinearities of the applied feed drive servomechanism. $G_n(s)$ represents the nominal plant model of the built-in velocity control loop and can be obtained using conventional identification methods [23].

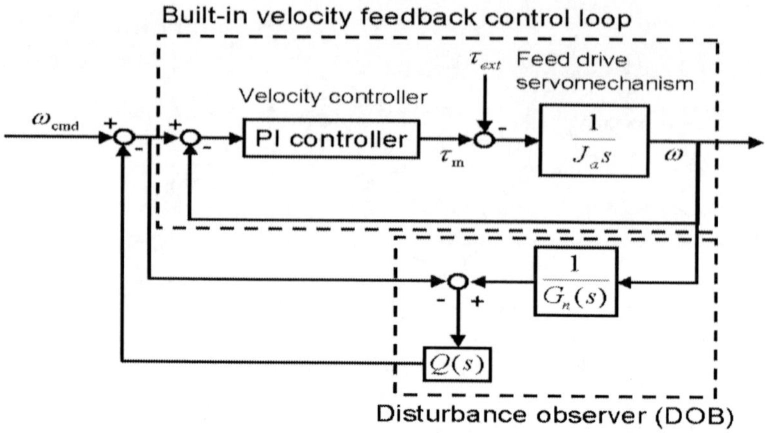

Figure 2.4. Modified velocity feedback control structure used in this chapter.

In this chapter, a built-in velocity control loop is formulated as a prototype second-order system, and the model parameters are obtained using an optimization method. The employed optimization method automatically searches model parameters to ensure that the velocity step responses of the model are similar to the actual velocity step responses of the built-in velocity control loop with different levels of velocity commands. $Q(s)$ is a low-pass filter used in the DOB feedback loop for disturbance compensation. The most important issue for the DOB is the design of this low-pass filter, $Q(s)$, because it significantly affects the stability and performance of the DOB feedback control system. Several papers have discussed the design of the low-pass filter, $Q(s)$, for a DOB structure [24-26]. However, without lost of generality, a simple first-order low-pass filter is considered in this chapter to simplify the implementation. The bandwidth of the filter is tuned manually to reduce the effects of disturbances containing high-frequency components and maintain the stability of the applied DOB feedback control system.

2.3. Experiments

As shown in Figure 2.1 and Figure 2.2, two experiments, the breakaway experiment [12,19] and constant-velocity experiment [12,17], were performed on a three-axis CNC milling machine, based on the measurement and control methods shown in Figure 2.3 and Figure 2.4. The two goals for these two experiments are listed below:

1. The breakaway experiment was employed to measure the minimum torque necessary to initiate motions of the moving part in the tested feed drive servomechanism. The experimental results of the breakaway experiment can also be used to identify the position-dependent friction of the tested feed drive servomechanism.
2. The constant-velocity experiment was employed to obtain friction and velocity data values in order to construct a friction-velocity map for the tested feed drive servomechanism. The experimental results of the constant-velocity experiment can also be used to identify the velocity-dependent friction of the tested feed drive servomechanism.

2.3.1. Breakaway Experiment

As suggested by Armstrong [12], the breakaway experiment is the simplest way to measure static friction, which is considered to be the position-dependent friction in this chapter, of a tested feed drive servomechanism. This section gives the details of an experiment that was performed with several tests. First, before the breakaway experiment, the applied servomotor pack was switched to a velocity control loop, and then the tested feed drive servomechanism was "warmed up" by a minute of motion through the workspace prior to collecting test data in order to eliminate the influence of the dwell time effect [12]. After this "warm up" procedure, the servomotor pack was set to the torque control loop such that torque commands could be directly used to actuate the tested feed drive servomechanism.

Second, the moving part of the tested feed drive servomechanism was positioned at a starting point and the torque command was gradually increased until the motor output torque could overcome the static friction of the tested feed drive servomechanism. Here, the torque command was increased from zero by sequentially adding step commands with a magnitude of 0.002 Nm per second. The motor output torque could be obtained from the interface signal of the amplifier of the applied servomotor pack [22]. As soon as the motor output torque overcame the static friction, the moving part moved, and the applied torque command was then brought to zero to allow the moving part to come to rest. The starting position (the position where the moving part starts its motion) and the breakaway torque (the motor output torque that overcame the static friction) were recorded during this test period. After the moving part came to rest, a new test period was initiated.

Numerous iterations were used in the breakaway experiment within the range of the home position to the target position, as shown in Figure 2.3. Moreover, ten breakaway experiments, each incorporating both the positive direction (from the home position to target position) and negative direction (from the target position to home position), were conducted to obtain the average behaviors of the frictions. Conventionally, it is difficult to obtain the breakaway torque in a test period by detecting the angular velocity of the applied servomotor at the moment that the motor output torque overcomes the static friction. The method proposed by Karnopp [15] was thus employed in this experiment. A small neighborhood around zero velocity, bounded by a threshold velocity, was defined in advance. If the angular velocity reached the threshold velocity, the measured torque value was used to represent the value of the static friction. In a case where the angular velocity was smaller than the threshold velocity, the velocity was considered to be zero when constructing the friction-velocity map of the tested feed drive servomechanism. When the angular velocity was larger than the threshold velocity, the measured torque was used to estimate the friction as a function of the velocity. In this chapter, because of the actuating limitation of the applied servomotor pack, the threshold velocity was set to 0.1571 rad/sec throughout the breakaway experiments.

Because the breakaway experiment can sample the static friction along a moving path with a high spatial sampling rate and the experimental results correlate very well with position-dependent perturbations [12], the breakaway experiment is usually employed to measure position-dependent frictions, and the experimental results are used to develop a friction compensation lookup table to compensate for these position-dependent frictions. Spectral analysis of the measured data values can also be used to analyze and identify the sources of the different position-dependent frictions [19].

2.3.2. Constant-Velocity Experiment

The constant-velocity experiment was used to obtain friction and velocity data values during the constant velocity movements of the tested feed drive servomechanism in order to construct a friction-velocity map to estimate the velocity-dependent frictions. As with the breakaway experiment, the applied servomotor pack was switched to a velocity control loop and a "warm up" procedure was performed on the tested feed drive servomechanism before the constant-velocity experiment. However, in comparison with the velocity PI control conventionally used in the constant-velocity experiment, the modified velocity feedback control structure shown in Figure 2.4 was used in this experiment to obtain good velocity tracking results and maintain constant velocity responses during the steady motions. The bandwidth of the modified velocity feedback control loop was 14.02 Hz.

In the constant-velocity experiment, as shown in Figure 2.3, the moving part of the tested feed drive servomechanism was positioned at return position A, and then a positive step velocity command was applied to the modified velocity feedback control loop to cause the moving part to move to return position B with a constant motion velocity. After that, a negative velocity step command with the same magnitude was applied to force the moving part to return to return position A. The velocity (obtained by angular position differentiation over a sampling period) and the motor output torque (obtained from the interface signal of the amplifier) were recorded every 10 msec during this test period. Ten iterations of the test period were conducted for a given step velocity command in order to obtain average data values, and thirteen step velocity commands with different levels ranging from 0.1571 rad/sec to 104.7198 rad/sec were given to obtain torque and velocity data values to construct a friction-velocity map. Moreover, a large number of data values was collected at a low velocity range in order to improve the identification of the nonlinear characteristics of the frictions. Ten iterations of the constant-velocity experiment were conducted in order to obtain the average behaviors of the frictions.

By applying the constant-velocity experiment to the tested feed drive servomechanism, a nominal velocity-dependent friction model was developed using the measured data values. The obtained nominal friction model can be used to analyze the velocity-dependent friction behaviors, and can also be applied to design the motion control of the feed drive servomechanism to compensate for these velocity-dependent frictions.

3. Friction Identification Techniques

The frictions of a feed drive servomechanism are usually difficult to estimate because they appear in a nonlinear fashion and the data values obtained by experiments are significantly influenced by unavoidable perturbations. Therefore, it is important to develop friction identification techniques for the tested feed drive servomechanism in order to obtain nominal friction models that can be used for compensation and control design purposes. This chapter considers Armstrong's findings that frictions are both position and velocity dependent [12]. The position-dependent friction is estimated by the breakaway experiment, while the velocity-dependent friction is modeled by using the data values obtained by the constant-velocity experiment.

3.1. Estimation of Position-Dependent Friction

Friction torques can be position-dependent in mechanical systems because of the imperfections in transmission systems. For instance, the imperfections induced by the misalignment of shaft and reductor centers could generate friction oscillations for industrial manipulators [27]. Although some experiments have further shown that the position-dependent friction is relatively weak [27], it is worth noting that the position-dependent friction becomes significant and cannot be neglected in other mechanical systems such as the feed drive servomechanisms in machine tools. In this chapter, the breakaway experiment described in sub-section 2.3 was applied to estimate the position-dependent friction of the X-axis and Y-axis feed drive servomechanisms, as shown in Figure 2.1 and Figure 2.2.

Figure 3.1 shows the experimental results of the breakaway experiment performed on the X-axis feed drive servomechanism without a payload. Here, the average values of the measured data are shown. These are used in the analysis of this chapter because they represent the repeatable part of the measured data and reduce the adverse effects induced by measurement noises. In order to investigate the positional dependency of the frictions, friction versus position curves in both the positive and negative directions are plotted in Figure 3.1. It was found out that these friction versus position curves vary significantly depending on the position of the applied servomotor, and thus the measured frictions are clearly position-dependent. Moreover, the position-dependent frictions not only exhibit oscillations with different frequencies but also show a linearly changed component that indicates that the friction changes linearly depending on the increment of the position. The misalignment imperfections in transmission systems generally induce oscillating frictions, while the out-of-parallel imperfection of the two support rails in a slide way mainly induces linearly changed frictions. The surface texture of the mechanical parts in the feed drive servomechanism also induces position-dependent frictions when one mechanical part contacts another to drive the moving part. Although the position-dependent frictions in the positive and negative directions show similar average magnitudes, their friction patterns are different because of the asymmetry effect of the frictions [12]. The asymmetry effect is a phenomenon that causes dissimilar friction behaviors for the different motion directions of a moving part. In this chapter, the asymmetry effect could be mainly induced by the changed contact surfaces of the two connected mechanical parts and the out-of-parallel imperfection of the two support rails. The contact surfaces of two connected mechanical parts such as the lead-screw bar and nut could change when the moving part moves in a different direction, and these changed contact surfaces could generate different friction behaviors. The out-of-parallel imperfection changes the contact force between two contact surfaces, and thus could generate different friction behaviors when the moving part moves in different directions. The standard deviation of the measured frictions is usually used to indicate their repeatability [12]. The average values of the standard deviation are 0.0055 Nm and 0.0062 Nm for the positive and negative directions, respectively, which are less than 5% of the average friction.

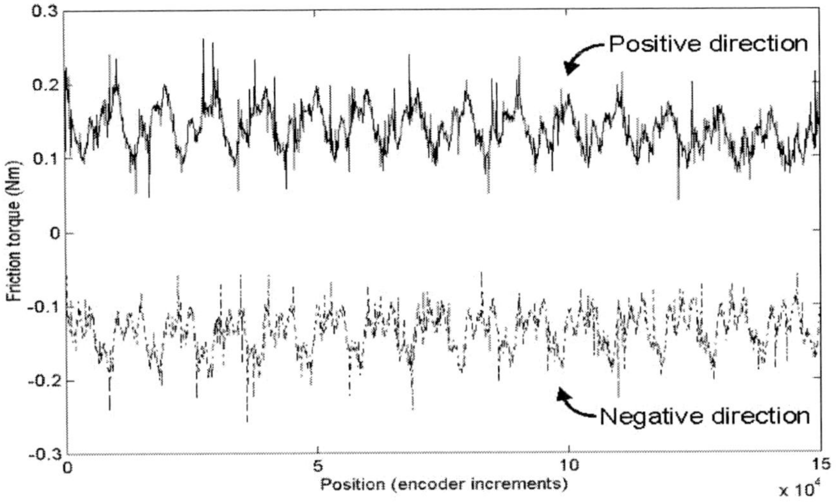

Figure 3.1. Position-dependent frictions obtained.

In this chapter, a spectral analysis using the FFT (fast Fourier transformation) algorithm [28] is performed to analyze the frequency characteristics of the position-dependent frictions obtained from the breakaway experiments. Figure 3.2 shows the FFT results for the obtained frictions shown in Figure 3.1. Here, the magnitude of each FFT component is normalized to determine its significance and show its percentage with respect to the average friction. A larger normalized magnitude indicates that the FFT component is more significant. The basic positional frequency used in the FFT analysis is 0.01 (encoder increments)$^{-1}$. This denotes that the static friction on the moving path from the home position to the target position, as shown in Figure 2.3, is sampled with a sampling period of 100 (encoder increments). More sampling instances are obtained during experiments where the moving path has a longer traveling distance and more refined FFT results are obtained when the basic positional frequency becomes higher. Although a higher basic frequency achieves more refined FFT results, a longer time is required to perform the experiment. In this chapter, the basic positional frequency is designed to be 0.01 (encoder increments)$^{-1}$ because the significant FFT components are found to be identical if the basic positional frequency is higher than 0.01 (encoder increments)$^{-1}$. Here, the FFT components with normalized magnitudes larger than 0.02 (2% of the zero FFT component, which denotes the average value of the measured frictions) are considered to be significant. Note that the FFT components are functions of the reciprocal values of the position increments. For instance, an FFT component with a positional frequency of "1/(1.0240×10^4)" corresponds to a friction that is repeatable and periodic with a positional period of 1.0240×10^4 (encoder increments); in other words, the friction repeats itself every 1.0240×10^4 encoder increments. Table 3.1 lists the significant FFT components shown in Figure 3.2. The FFT component with positional frequency 0 (encoder increments)$^{-1}$ represents the average static friction of the tested feed drive servomechanism. Because one revolution of the motor shaft is equivalent to 10,000 encoder increments and the lead-screw bar is connected to the motor shaft through a coupler, the FFT component with a positional frequency near 9.7656×10^{-5} (encoder increments)$^{-1}$ can be considered the static friction generated from the interaction between the motor shaft and lead-screw bar of the feed drive servomechanism. As shown in Table 3.1, the other significant positional frequencies are the frequencies close to the harmonic frequencies of the FFT component with a positional frequency near 9.7656×10^{-5} (encoder increments)$^{-1}$. Hence, the other significant FFT components are also generated from the interaction between the motor shaft and lead-screw bar. The other FFT components could be considered static friction generated from the interaction between the moving part and the support rails of the slide way used in the tested feed drive servomechanism. Usually, it is difficult to identify the sources of the other FFT components because they are significantly affected by the surface textures of the moving part and support rails. Moreover, different manufacturing processes usually generate different surface textures for machined parts. Mechanical parts with small and regular distributed surface structures can cause FFT components with normalized magnitudes of less than 0.02. The random friction component exists because of the establishment and breaking of asperity contacts between sliding surfaces [19].

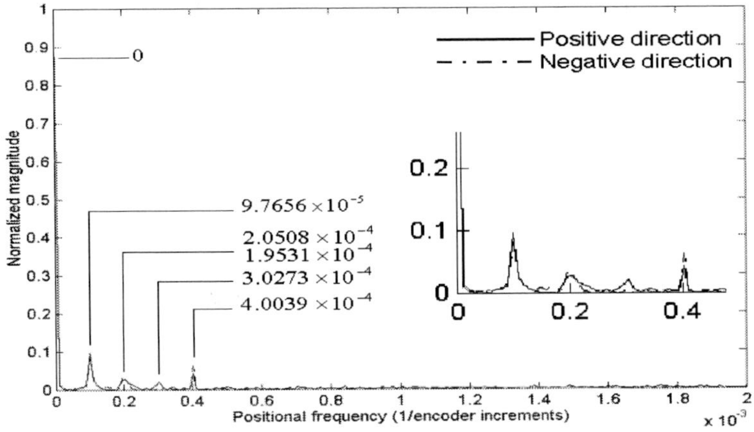

Figure 3.2. FFT results for obtained position-dependent frictions.

Table 3.1. Significant components of FFT analysis

No.	Positive direction		
	Positional frequency ((encoder increments)$^{-1}$)	Positional period (encoder increments)	Normalized magnitude
1	0	—	1
2	9.7656×10^{-5}	1.0240×10^{4}	9.6094×10^{-2}
3	2.0508×10^{-4}	4.8762×10^{3}	2.9107×10^{-2}
4	3.0273×10^{-4}	3.3032×10^{3}	2.1176×10^{-2}
5	4.0039×10^{-4}	2.4976×10^{3}	4.2183×10^{-2}
No.	Negative direction		
	Positional frequency ((encoder increments)$^{-1}$)	Positional period (encoder increments)	Normalized magnitude
1	0	—	1
2	9.7656×10^{-5}	1.0240×10^{4}	8.3959×10^{-2}
3	1.9531×10^{-4}	5.120×10^{3}	3.5844×10^{-2}
4	—	—	—
5	4.0039×10^{-4}	2.4976×10^{3}	6.4360×10^{-2}

The measured static frictions, which contain a random static friction component as shown in Figure 3.1 and Figure 3.2, cannot be directly applied to the development of a friction compensation lookup table. Therefore, a low-pass filter is applied to filter out the random static friction component in this chapter. In order to preserve the significant components of the static friction, as shown in Table 3.1, the bandwidth of the low-pass filter is designed to be 0.0005 (encoder increments)$^{-1}$.

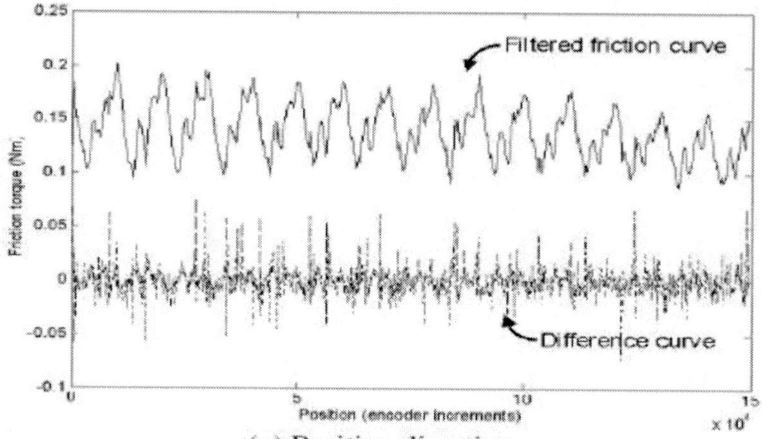

(a) Positive direction

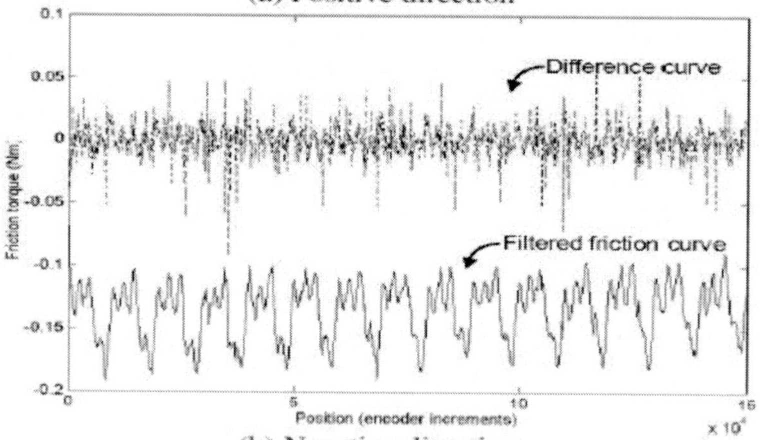

(b) Negative direction

Figure 3.3. Filtered static friction curves.

Figure 3.3 shows the filtered static friction curves in both the positive and negative directions. Moreover, the filtered static friction was subtracted from the measured static friction in order to show the validity of the filtered static friction curve. Figure 3.3 also shows the differences obtained in the two different directions. For the difference in the positive direction, the calculated average value is 2.1239×10^{-4} Nm and the value of standard deviation is 1.3238×10^{-2} Nm. For the difference in the negative direction, the calculated average value is -1.8010×10^{-4} Nm and the value of standard deviation is

1.2479×10^{-2} Nm. According to the statistical results, it is feasible to apply the filtered static friction curves of Figure 3.3 to the development of friction compensation lookup tables because they describe most of the repeatable parts of the static frictions and filter out random frictions. The differences in both cases have average values almost equal to zero. The standard deviation values of the differences in both cases are lower than 10% of the corresponding average static frictions.

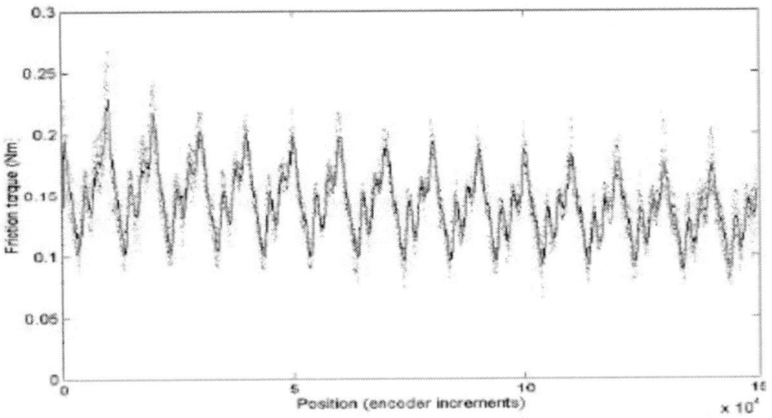

(a) Positive direction

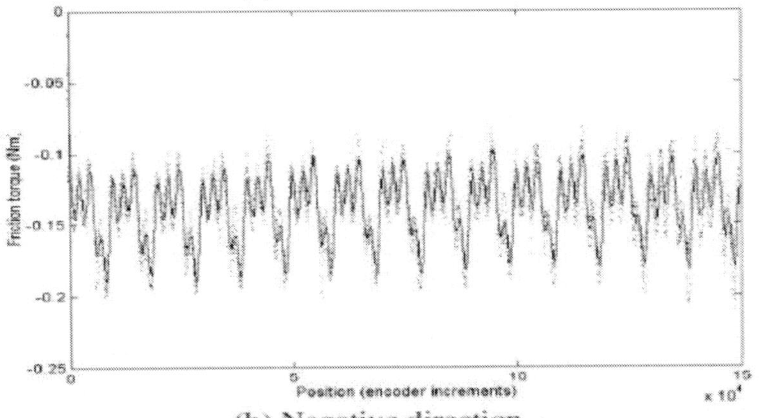

(b) Negative direction

Figure 3.4. Average and filtered static friction curves for different loading conditions.

Because different loading conditions could lead to different static friction characteristics in feed drive servomechanisms, a friction compensation lookup table must consider the static frictions caused by different loading conditions. In this chapter, the payload of the tested feed drive servomechanism was modified by adding weights in increments of 1 kg. Thus, different loading conditions were obtained from no load to a 5 kg maximum payload. Further, the filtered static friction curves for different loading conditions were obtained, as shown in Figure 3.4, by using the friction processing methods previously mentioned. The average static friction curve obtained by averaging the static frictions, as shown in Figure 3.4, was then used to develop friction compensation lookup tables for the different motion directions.

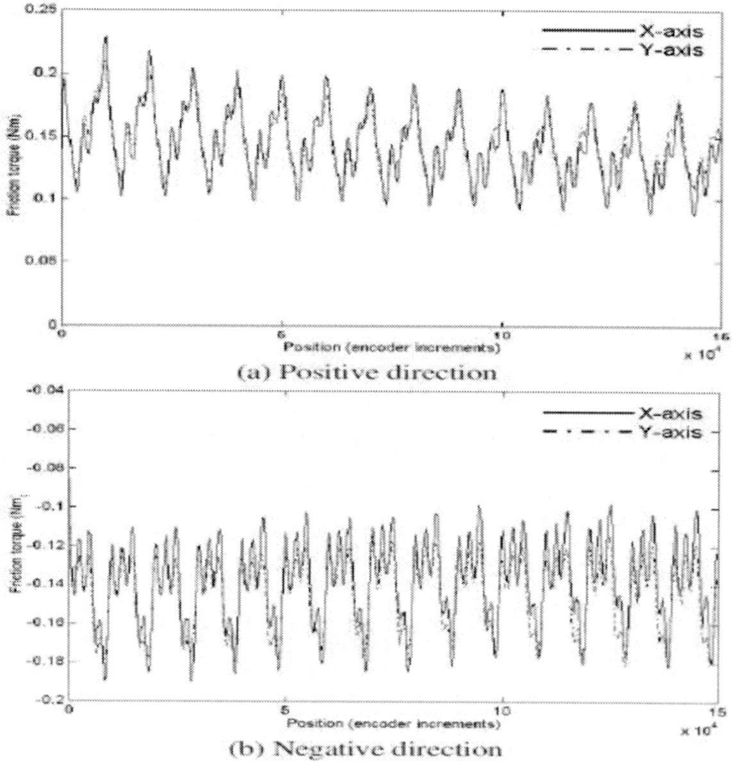

Figure 3.5. Friction compensation lookup tables. (a) Positive direction. (b) Negative direction.

By applying the same approaches, the friction compensation lookup tables for the Y-axis feed drive servomechanism in the experimental setup were obtained, as shown in Figure 3.5. Figure 3.5 also shows the friction compensation lookup tables for the X-axis feed drive servomechanism. In this chapter, the friction compensation lookup tables shown in Figure 3.5 are used to compensate for the static frictions of the X-axis and Y-axis feed drive servomechanisms in order to improve the motion accuracy of the applied three-axis CNC milling machine.

3.2. Estimation of Velocity-Dependent Friction

Velocity-dependent friction is usually modeled as a map between the velocity and friction torque, which is mainly composed of Coulomb friction, viscous friction, and the other retarding torques closely related to the magnitude and direction of the velocity [29]. The constant-velocity experiment is the usual approach employed to model the velocity-dependent friction. However, in this chapter, the results of the constant-velocity experiment described in sub-section 2.3 are employed to obtain friction and velocity data values in order to construct friction-velocity maps for the estimation of the velocity-dependent frictions of the X-axis and Y-axis feed drive servomechanisms, as shown in Figure 2.1 and Figure 2.2.

Consider the feed drive servomechanism, actuated by a servomotor, with an equation of motion as

$$\tau_m - \tau_f = J_a\dot{\omega} = J_a\ddot{\theta} \tag{3.1}$$

where J_a denotes the equivalent inertia of the feed drive servomechanism; θ and ω are the motor shaft angular position and angular velocity, respectively; $\dot{\omega}$ (or $\ddot{\theta}$) denotes the angular acceleration of the motor shaft; τ_m is the motor output torque; and τ_f is the friction torque. The equivalent inertia, J_a, is the combination of motor inertia (the inertia of the rotor) and load inertia (the compound inertia of the coupler, lead-screw bar, worktable, and payload). The motor shaft angular position, θ, is measured by means of an optical encoder with a resolution of 10,000 encoder increments/revolutions or 6.2832×10^{-4} radian/encoder increments, and the motor shaft angular velocity, ω, is computed by angular position differentiation over a sampling period. The motor output torque, τ_m, is well controlled by an inner current loop with a PI feedback controller such that the motor output torque could rapidly follow a torque command with small steady errors. The friction torque, τ_f, includes both position-dependent friction, τ_f^θ, and velocity-dependent friction, τ_f^ω. Therefore, Eq. (3.1) can be further rewritten as

$$\tau_m - (\tau_f^\theta + \tau_f^\omega) = J_a\dot{\omega} = J_a\ddot{\theta} \tag{3.2}$$

If the velocity response of the tested feed drive servomechanism is a constant, ω_s, during steady motions, i.e., $\omega = \omega_s$ and $\dot{\omega} = 0$ at the steady state, Eq. (3.2) can be rewritten as

$$\tau_m - (\tau_f^\theta + \tau_f^\omega) = 0 \tag{3.3}$$

and thus

$$\tau_m = \tau_f^\theta + \tau_f^\omega = \tau_f \tag{3.4}$$

Therefore, the friction torque, τ_f, can be obtained by measuring the motor output torque, τ_{m}, during steady motions. By collecting different levels of friction and velocity data values, τ_f and ω_s, through constant-velocity experiments, a friction-velocity map is constructed for modeling the velocity-dependent friction, τ_f^ω. However, Eq. (3.3) shows that the velocity response, ω_s, must be kept constant during steady motions, and Eq. (3.4) shows that the measured friction torque, τ_f, is contaminated by the position-dependent friction, τ_f^θ. The friction and velocity data values obtained by applying conventional constant-velocity experiments could deteriorate the friction-velocity map and thus affect the validity of the obtained velocity-dependent friction model. Therefore, a new data extraction method is developed in this chapter that removes the effects induced by position-dependent friction to obtain reliable friction and velocity data values. This data extraction method is also used to construct a new kind of friction-velocity map. Moreover, the results of the constant-velocity experiment described in sub-section 2.3 are used to obtain raw friction and velocity data values.

In this chapter, as shown in Figure 2.4, the modified velocity feedback control structure is applied to keep the velocity responses of the tested feed drive servomechanism constant during steady motions in the constant-velocity experiment. The nominal plant of the built-in velocity control loop is obtained as

$$G_n(s) = \frac{16978.09}{s^2 + 158.966s + 16978.09} \tag{3.5}$$

Figure 3.6 shows the step responses of the built-in velocity control loop and the nominal plant, $G_n(s)$, with different velocity command levels. The experimental results show that the nominal plant in Eq. (3.5) provides a good approximation to the built-in velocity control loop. Moreover, the filter, $Q(s)$, in Figure 2.4 is designed as a low-pass filter with unity gain and a bandwidth of 15.92 Hz. Figure 3.7 shows the steady responses of the velocity feedback control systems with and without using a disturbance observer. For the system without a DOB, the average value of the steady response is 5.1451 rad/sec, and the variance value of the steady response is 1.4108×10^{-2} rad/sec. However, for a system with a DOB, the average value of the steady response becomes 5.2356 rad/sec, and the variance value of the steady response becomes 1.4750×10^{-3} rad/sec. Clearly, the modified velocity feedback control structure causes the tested feed drive servomechanism to provide steadier motions than that with a built-in velocity control structure that only uses velocity PI feedback control.

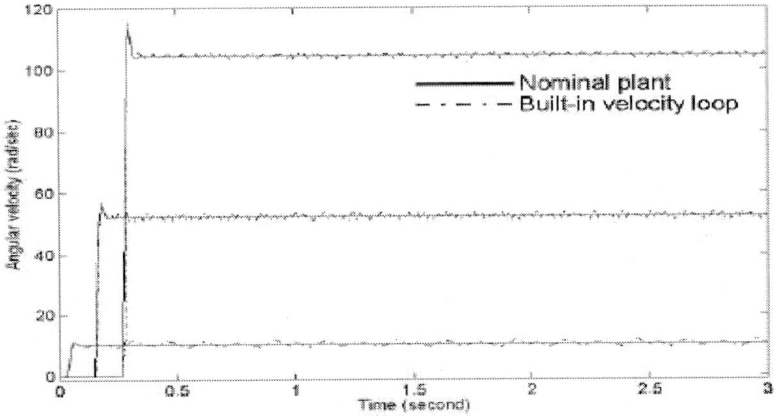

Figure 3.6. Step responses of built-in velocity control loop and nominal plant.

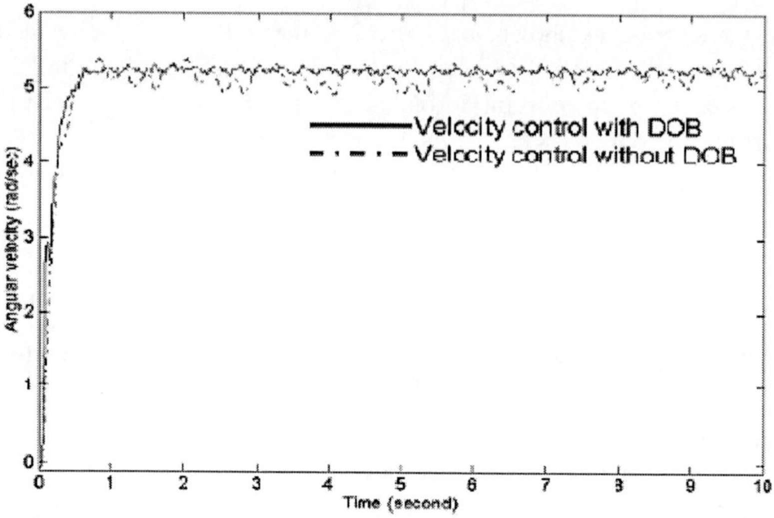

Figure 3.7. Steady responses of velocity feedback control systems with and without DOB.

Because the modified velocity feedback control structure makes the velocity response, ω_s, almost constant during steady motions, i.e., $\dot{\omega}_s \cong 0$, the equation of motion at the steady state can be obtained as

$$\tau_m - (\tau_f^{\theta} + \tau_f^{\omega_s}) = J_a \dot{\omega}_s \cong 0 \tag{3.6}$$

where $\tau_f^{\omega_s}$ denotes the velocity-dependent friction corresponding to the angular velocity, ω_s. Consider a tested feed drive servomechanism actuated by motor output torques τ_{m1} and τ_{m2} at the same angular position, θ. The velocity response approaches ω_{s1} and ω_{s2} at position θ corresponding to the motor output torques, τ_{m1} and τ_{m2}, respectively. Because the position-dependent friction, τ_f^{θ}, is invariant at position θ, Eq. (3.6) can be used to obtain the equation of motion at position θ with different motor output torques as

$$\tau_{m1} - \tau_f^{\omega_{s1}} - \tau_f^{\theta} \cong 0 \tag{3.7}$$

and

$$\tau_{m2} - \tau_f^{\omega_{s2}} - \tau_f^{\theta} \cong 0 \tag{3.8}$$

where $\tau_f^{\omega_{s1}}$ and $\tau_f^{\omega_{s2}}$ denote the velocity-dependent frictions at position θ corresponding to angular velocities ω_{s1} and ω_{s2}, respectively. By subtracting Eq. (3.7) from Eq. (3.8), the torque difference equation can be obtained as

$$(\tau_{m2} - \tau_{m1}) - (\tau_f^{\omega_{s2}} - \tau_f^{\omega_{s1}}) \cong 0 \tag{3.9}$$

In this chapter, a polynomial friction model [12] without the static friction term is applied to illustrate the velocity-dependent friction of the tested feed drive servomechanism as

$$\tau_f^{\omega} = \sum_{i=1}^{m} a_i \omega^i \tag{3.10}$$

where m denotes the order of the applied polynomial function. Therefore, the friction difference, $\tau_f^{\omega_{s2}} - \tau_f^{\omega_{s1}}$, can be obtained as

$$\tau_f^{\omega_{s2}} - \tau_f^{\omega_{s1}} = \sum_{i=1}^{m} a_i \omega_{s2}^i - \sum_{i=1}^{m} a_i \omega_{s1}^i = \sum_{i=1}^{m} a_i (\omega_{s2}^i - \omega_{s1}^i) \tag{3.11}$$

By substituting Eq. (3.11), the torque difference equation, Eq. (3.9), becomes

$$\begin{aligned}(\tau_{m2} - \tau_{m1}) &= \sum_{i=1}^{m} a_i (\omega_{s2}^i - \omega_{s1}^i) \\ &= \begin{bmatrix}(\omega_{s2}^1 - \omega_{s1}^1) & (\omega_{s2}^2 - \omega_{s1}^2) & \cdots & (\omega_{s2}^m - \omega_{s1}^m)\end{bmatrix} \cdot \begin{bmatrix} a_1 \\ a_2 \\ \vdots \\ a_m \end{bmatrix}\end{aligned} \tag{3.12}$$

We can define data set (τ_{mb}, ω_{sb}), composed of motor output torque τ_{mb} and angular velocity ω_{sb}, to be the base data set, and data set (τ_{mj}, ω_{sj}), composed of motor output torque τ_{mj} and angular velocity ω_{sj}, to be the target data set for $j = 1, \ldots, \mathrm{n}$. Then, the torque difference equation, Eq. (3.12), can be rewritten as

$$(\tau_{mj} - \tau_{mb}) = \sum_{i=1}^{m} a_i(\omega_{sj}^i - \omega_{sb}^i)$$
$$= \begin{bmatrix}(\omega_{sj}^1 - \omega_{sb}^1) & (\omega_{sj}^2 - \omega_{sb}^2) & \cdots & (\omega_{sj}^m - \omega_{sb}^m)\end{bmatrix} \cdot \begin{bmatrix} a_1 \\ a_2 \\ \vdots \\ a_m \end{bmatrix} \tag{3.13}$$

for $j = 1, \ldots, \mathrm{n}$. Further, by combining all of the data sets, the torque difference equation, Eq. (3.13), can be extended as

$$\begin{bmatrix} \tau_{m1} - \tau_{mb} \\ \tau_{m2} - \tau_{mb} \\ \vdots \\ \tau_{mn} - \tau_{mb} \end{bmatrix} = \begin{bmatrix} (\omega_{s1}^1 - \omega_{sb}^1) & (\omega_{s1}^2 - \omega_{sb}^2) & \cdots & (\omega_{s1}^m - \omega_{sb}^m) \\ (\omega_{s2}^1 - \omega_{sb}^1) & (\omega_{s2}^2 - \omega_{sb}^2) & \cdots & (\omega_{s2}^m - \omega_{sb}^m) \\ \vdots & \vdots & \ddots & \vdots \\ (\omega_{sn}^1 - \omega_{sb}^1) & (\omega_{sn}^2 - \omega_{sb}^2) & \cdots & (\omega_{sn}^m - \omega_{sb}^m) \end{bmatrix} \begin{bmatrix} a_1 \\ a_2 \\ \vdots \\ a_m \end{bmatrix} \tag{3.14}$$

Here, the values of all of the data sets, including the base data set, (τ_{mb}, ω_{sb}), and target data sets, (τ_{mj}, ω_{sj}), for $j = 1, \ldots, \mathrm{n}$, can be obtained by the constant-velocity experiment described in sub-section 2.3. Moreover, an optimization algorithm must be applied to compute the model parameters $\begin{bmatrix} a_1 & a_2 & \cdots & a_m \end{bmatrix}^T$ to minimize estimation error ε as

$$\varepsilon = \left\| \begin{bmatrix} \tau_{\mathrm{m1}} - \tau_{\mathrm{mb}} \\ \tau_{\mathrm{m2}} - \tau_{\mathrm{mb}} \\ \vdots \\ \tau_{\mathrm{mn}} - \tau_{\mathrm{mb}} \end{bmatrix} - \begin{bmatrix} (\omega_{\mathrm{s1}}^1 - \omega_{\mathrm{sb}}^1) & (\omega_{\mathrm{s1}}^2 - \omega_{\mathrm{sb}}^2) & \cdots & (\omega_{\mathrm{s1}}^{\mathrm{m}} - \omega_{\mathrm{sb}}^{\mathrm{m}}) \\ (\omega_{\mathrm{s2}}^1 - \omega_{\mathrm{sb}}^1) & (\omega_{\mathrm{s2}}^2 - \omega_{\mathrm{sb}}^2) & \cdots & (\omega_{\mathrm{s2}}^{\mathrm{m}} - \omega_{\mathrm{sb}}^{\mathrm{m}}) \\ \vdots & \vdots & \ddots & \vdots \\ (\omega_{\mathrm{sn}}^1 - \omega_{\mathrm{sb}}^1) & (\omega_{\mathrm{sn}}^2 - \omega_{\mathrm{sb}}^2) & \cdots & (\omega_{\mathrm{sn}}^{\mathrm{m}} - \omega_{\mathrm{sb}}^{\mathrm{m}}) \end{bmatrix} \begin{bmatrix} \mathrm{a}_1 \\ \mathrm{a}_2 \\ \vdots \\ \mathrm{a}_{\mathrm{m}} \end{bmatrix} \right\| \tag{3.15}$$

In this chapter, the base data set (τ_{mb}, ω_{sb}) is recorded when velocity response ω_{sb} is equal to the threshold velocity in the breakaway experiment described in sub-section 2.3. Thus, the velocity-dependent friction is set to zero when velocity response ω is less than velocity response ω_{sb}. Therefore, the velocity-dependent friction model in this chapter becomes

$$\hat{\tau}_f^{\omega} = \begin{cases} 0 & ,0 < \omega < \omega_{sb} \\ \sum_{i=1}^{m} a_i \omega^i , \omega \geq \omega_{sb} > 0 \end{cases} \quad (3.16)$$

where ω is the motor shaft angular velocity and $\hat{\tau}_f^{\omega}$ denotes the estimation and computation result of the velocity-dependent friction model in Eq. (3.16). Here, Eq. (3.10) to Eq. (3.16) can also be used to model the velocity-dependent friction when the moving part of the tested feed drive servomechanism moves in the negative direction. By considering the velocity-dependent friction in the negative motion direction, the velocity-dependent friction model is modified to

$$\hat{\tau}_f^{\omega} = \begin{cases} \sum_{i=1}^{m\_neg} a_i^{neg} \omega^i , \omega \leq -\omega_{sb}^{neg} \\ 0, -\omega_{sb}^{neg} < \omega < \omega_{sb}^{pos} \\ \sum_{i=1}^{m\_pos} a_i^{pos} \omega^i , \omega \geq \omega_{sb}^{pos} \end{cases} \quad (3.17)$$

where m_neg and m_pos denote the order of the friction model in the negative and positive motion directions, respectively. ω_{sb}^{neg} and ω_{sb}^{pos} are the base velocity responses in the negative and positive motion directions, respectively. The coefficient $\begin{bmatrix} a_1^{neg} & a_2^{neg} & \cdots & a_{m\_neg}^{neg} \end{bmatrix}^T$ represents the model parameters in the negative motion direction and is usually different from coefficient $\begin{bmatrix} a_1^{pos} & a_2^{pos} & \cdots & a_{m\_pos}^{pos} \end{bmatrix}^T$, which represents the model parameters in the positive motion direction.

Therefore, we can summarize the procedure for modeling the velocity-dependent friction of the tested feed drive servomechanism as follows:

(a) Obtain the base data sets, $(\tau_{mb}^{pos}, \omega_{sb}^{pos})$ and $(\tau_{mb}^{neg}, \omega_{sb}^{neg})$, and the target data sets, $(\tau_{mj}^{pos}, \omega_{sj}^{pos})$ and $(\tau_{mj}^{neg}, \omega_{sj}^{neg})$, for $j = 1, \ldots, \mathrm{n}$ using the results of the constant-velocity experiment described in sub-section 2.3.

(b) Sort the values of these obtained data sets by angular position ranging from the home position to the target position, as shown in Figure 2.3.

(c) Set the order of the friction model in the positive motion direction, m_pos.

(d) Compute the torque difference, $(\tau_{mj}^{pos} - \tau_{mb}^{pos})$, for $j = 1, \ldots, \mathrm{n}$ using Eq. (3.13) and construct the extended equation as Eq. (3.14).

(e) Compute the parameters, $\begin{bmatrix} a_1^{pos} & a_2^{pos} & \cdots & a_{m\_pos}^{pos} \end{bmatrix}^T$, that minimize the estimation error, ε, as in Eq. (3.15).

(f) Set the order of the friction model in the negative motion direction, m_neg.

(g) Compute the torque difference, $(\tau_{mj}^{neg} - \tau_{mb}^{neg})$, for $j = 1, \ldots, \mathrm{n}$ and construct the extended equation.

(h) Compute the parameters, $\begin{bmatrix} a_1^{neg} & a_2^{neg} & \cdots & a_{m\_neg}^{neg} \end{bmatrix}^T$, that minimize the estimation error.

The friction model shown in Eq. (3.18) [18,29,30] can be applied to describe the friction characteristics of the applied feed drive servomechanisms.

$$\tau_f^{\omega_s} = (\alpha_0 + \alpha_1 e^{-(\frac{\omega_s}{\omega_0})^2})\mathrm{sgn}(\omega_s) + \alpha_2 \omega_s \tag{3.18}$$

where $\tau_f^{\omega_s}$ is the friction torque corresponding to the steady-state velocity, ω_s. $(\alpha_0 + \alpha_1)$ is the static friction and α_2 represents the viscous friction. ω_0 denotes the Stribeck velocity. The friction difference, $\tau_f^{\omega_{s2}} - \tau_f^{\omega_{s1}}$, can be obtained as

$$\begin{aligned}&\tau_f^{\omega_{s2}} - \tau_f^{\omega_{s1}} \\ &= \left[(\alpha_0 + \alpha_1 e^{-(\frac{\omega_{s2}}{\omega_0})^2})\mathrm{sgn}(\omega_{s2}) + \alpha_2\omega_{s2}\right] - \left[(\alpha_0 + \alpha_1 e^{-(\frac{\omega_{s1}}{\omega_0})^2})\mathrm{sgn}(\omega_{s1}) + \alpha_2\omega_{s1}\right] \\ &= \begin{cases} \alpha_1(e^{-(\frac{\omega_{s2}}{\omega_0})^2} - e^{-(\frac{\omega_{s1}}{\omega_0})^2}) + \alpha_2(\omega_{s2} - \omega_{s1}), \text{ for } \omega_{s2} > 0 \text{ and } \omega_{s1} > 0 \\ \alpha_1[-(e^{-(\frac{\omega_{s2}}{\omega_0})^2} - e^{-(\frac{\omega_{s1}}{\omega_0})^2})] + \alpha_2(\omega_{s2} - \omega_{s1}), \text{ for } \omega_{s2} < 0 \text{ and } \omega_{s1} < 0 \end{cases}\end{aligned} \quad (3.19)$$

By substituting Eq. (3.19), the torque difference equation, Eq. (3.9), becomes

$$\begin{aligned}&(\tau_{m2} - \tau_{m1}) = \tau_f^{\omega_{s2}} - \tau_f^{\omega_{s1}} \\ &= \begin{cases} \alpha_1(e^{-(\frac{\omega_{s2}}{\omega_0})^2} - e^{-(\frac{\omega_{s1}}{\omega_0})^2}) + \alpha_2(\omega_{s2} - \omega_{s1}), \text{ for } \omega_{s2} > 0 \text{ and } \omega_{s1} > 0 \\ \alpha_1[-(e^{-(\frac{\omega_{s2}}{\omega_0})^2} - e^{-(\frac{\omega_{s1}}{\omega_0})^2})] + \alpha_2(\omega_{s2} - \omega_{s1}), \text{ for } \omega_{s2} < 0 \text{ and } \omega_{s1} < 0 \end{cases}\end{aligned} \quad (3.20)$$

Define data set (τ_{mb}, ω_{sb}), composed of motor output torque τ_{mb} and angular velocity ω_{sb}, to be the base data set and data set (τ_{mj}, ω_{sj}), composed of motor output torque τ_{mj} and angular velocity ω_{sj}, to be the target data set for $j = 1, \ldots, \mathrm{n}$. Then, the torque difference equation, Eq. (3.20), can be rewritten as

$$\begin{aligned}&(\tau_{mj} - \tau_{mb}) = \tau_f^{\omega_{sj}} - \tau_f^{\omega_{sb}} \\ &= \begin{cases} \alpha_1(e^{-(\frac{\omega_{sj}}{\omega_0})^2} - e^{-(\frac{\omega_{sb}}{\omega_0})^2}) + \alpha_2(\omega_{sj} - \omega_{sb}), \text{ for } \omega_{sj} > 0 \text{ and } \omega_{sb} > 0 \\ \alpha_1[-(e^{-(\frac{\omega_{sj}}{\omega_0})^2} - e^{-(\frac{\omega_{sb}}{\omega_0})^2})] + \alpha_2(\omega_{sj} - \omega_{sb}), \text{ for } \omega_{sj} < 0 \text{ and } \omega_{sb} < 0 \end{cases}\end{aligned} \quad (3.21)$$

for $j = 1, \ldots, \mathrm{n}$.

Here, an optimization algorithm must be applied to compute the model parameters $(\alpha_1, \alpha_2, \omega_0)$ to minimize the estimation error, ε, as

$$\varepsilon = \sqrt{\sum_{j=1}^{n}\left[(\tau_{mj}-\tau_{mb})-(\tau_f^{\omega_{sj}}-\tau_f^{\omega_{sb}})\right]^2} \qquad (3.22)$$

In this chapter, the *fminu* optimization function provided by the Matlab optimization toolbox [31] is used to obtain the model parameters $(\alpha_1,\alpha_2,\omega_0)$. The base data set (τ_{mb},ω_{sb}) is recorded when velocity response ω_{sb} is equal to the smallest velocity in the constant-velocity experiment described in sub-section 2.3. Because of the actuating limitation of the applied servomotor pack, the smallest velocity is set to be 0.1571 rad/sec. Therefore, we can summarize the procedure for modeling the velocity-dependent friction of the tested feed drive servomechanism as follows:

(a) Obtain the static frictions, τ_e^{neg} and τ_e^{pos}, using the results of the breakaway experiment described in sub-section 2.3.
(b) Obtain the base data sets, $(\tau_{mb}^{pos},\omega_{sb}^{pos})$ and $(\tau_{mb}^{neg},\omega_{sb}^{neg})$, and the target data sets, $(\tau_{mj}^{pos},\omega_{sj}^{pos})$ and $(\tau_{mj}^{neg},\omega_{sj}^{neg})$, for $j=1,\ldots,\mathrm{n}$ using the results of the constant-velocity experiment described in sub-section 2.3.
(c) Sort the values of these obtained data sets by angular position ranging from the home position to the target position, as shown in Figure 2.3.
(d) Compute the torque difference, $(\tau_{mj}^{pos}-\tau_{mb}^{pos})$, for $j=1,\ldots,\mathrm{n}$.
(e) Compute the parameters $(\alpha_1^{pos},\alpha_2^{pos},\omega_0^{pos})$ using an optimization algorithm.
(f) Compute the torque difference, $(\tau_{mj}^{neg}-\tau_{mb}^{neg})$, for $j=1,\ldots,\mathrm{n}$.
(g) Compute the parameters $(\alpha_1^{neg},\alpha_2^{neg},\omega_0^{neg})$ using an optimization algorithm.

Figure 3.8 shows the measured motor output torques and the corresponding angular velocities of a constant-velocity experiment performed on the X-axis feed drive servomechanism without a payload.

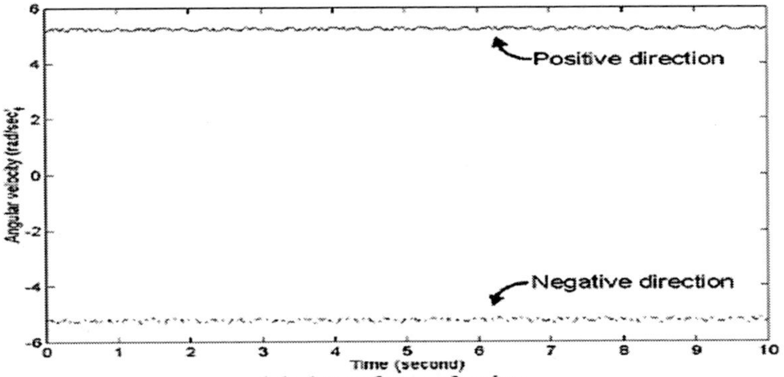

(a) Angular velocity

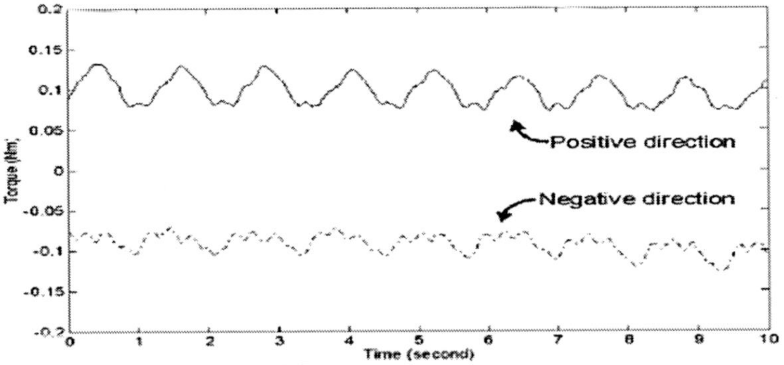

(b) Motor output torque

Figure 3.8. Data measured by constant-velocity experiment.

The motor output torque was found to fluctuate obviously, indicating that the motor output torque is position dependent. In other words, the position-dependent friction significantly affects the motor output torque, and the previously mentioned friction extraction method is thus applied to extract velocity-dependent frictions.

In order to generate the base data sets, $(\tau_{mb}^{pos}, \omega_{sb}^{pos})$ and $(\tau_{mb}^{neg}, \omega_{sb}^{neg})$, and the target data sets, $(\tau_{mj}^{pos}, \omega_{sj}^{pos})$ and $(\tau_{mj}^{neg}, \omega_{sj}^{neg})$, for $j = 1, \ldots, n$, the motor output torque versus position curves and the steady-state velocity versus position curves that belong to the same set of experiments are obtained, as shown in Figure 3.9. Here, in order to represent the repeatable part of the measured data and reduce the effects induced by measurement noise, the average values of the measured data are shown in Figure 3.9 and used for further computations.

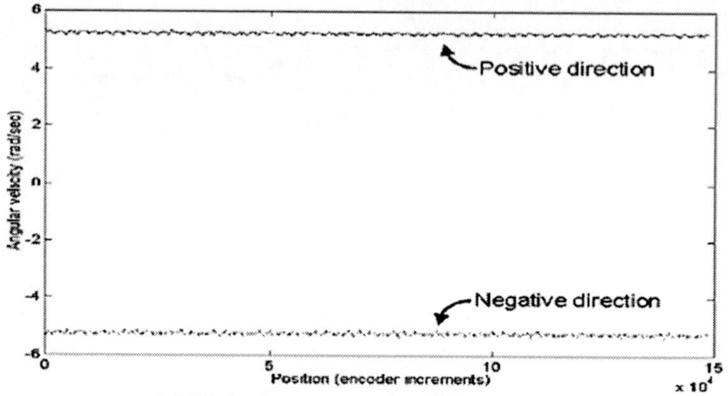

(a) Velocity versus position curves

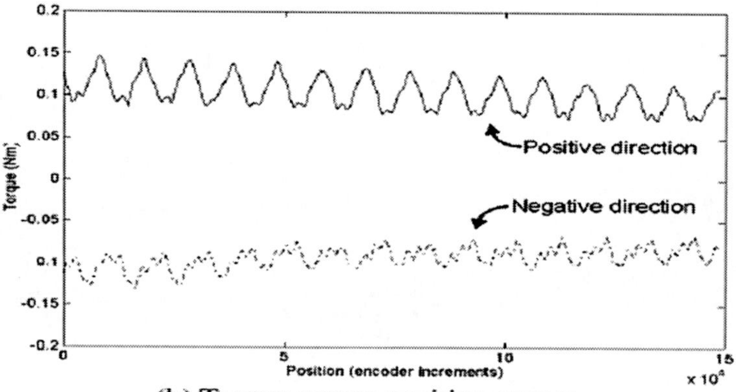

(b) Torque versus position curves

Figure 3.9. Torque versus position curves and velocity versus position curves.

The averaged values of the torque differences, $(\tau_{mj}^{pos} - \tau_{mb}^{pos})$ and $(\tau_{mj}^{neg} - \tau_{mb}^{neg})$, for $j = 1, \ldots, \text{n}$ were used to construct the extended equation shown in Eq. (3.14). In this chapter, ten iterations of the constant-velocity experiment were conducted to obtain the average behaviors of the frictions. Figure 3.10 shows the average values and the values of the standard deviation of the torque differences at different velocities using the data obtained from the ten constant-velocity experiments. The standard deviation values for all of the different velocities are lower than 10% of the corresponding average values.

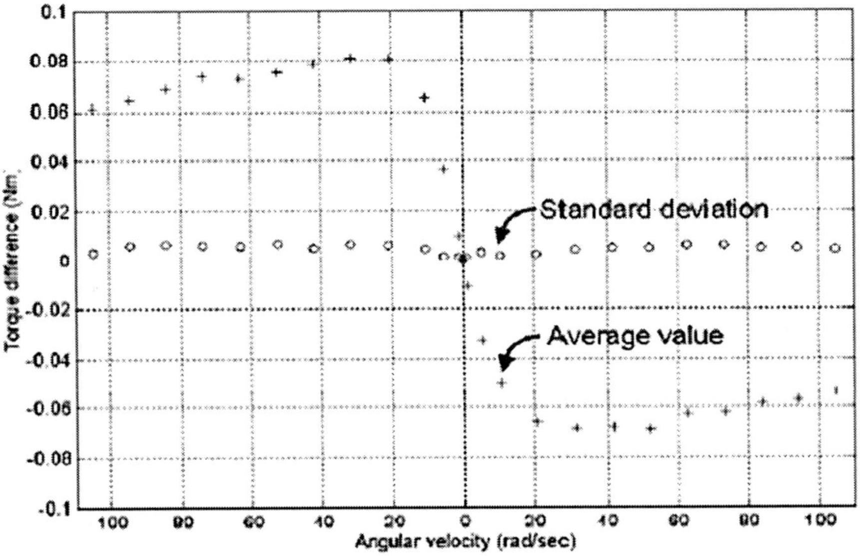

Figure 3.10. Average values and values of standard deviation of torque differences.

Similar to the development of the friction compensation lookup tables, the development of a velocity-dependent friction model must consider the friction characteristics caused by different loading conditions. Thus, different loading conditions for the tested feed drive servomechanism were employed in this chapter, from no load to a 5 kg maximum payload, by adding weights in 1-kg increments. Figure 3.11 shows the average torque difference values under different loading conditions found using the previously mentioned data processing methods. The average torque differences obtained by averaging the average torque difference values under different loading conditions, as shown in Figure 3.11, were then used to construct the extended equation shown in Eq. (3.14) for further computing the model parameters, $\begin{bmatrix} a_1^{pos} & a_2^{pos} & \cdots & a_{m\_pos}^{pos} \end{bmatrix}^T$ in the positive direction and $\begin{bmatrix} a_1^{neg} & a_2^{neg} & \cdots & a_{m\_neg}^{neg} \end{bmatrix}^T$ in the negative direction. Table 3.2 lists the parameters obtained for the friction model, as shown in Eq. (3.17), and the corresponding estimation errors found using a least-square optimization method. Figure 3.12 shows the modeling results.

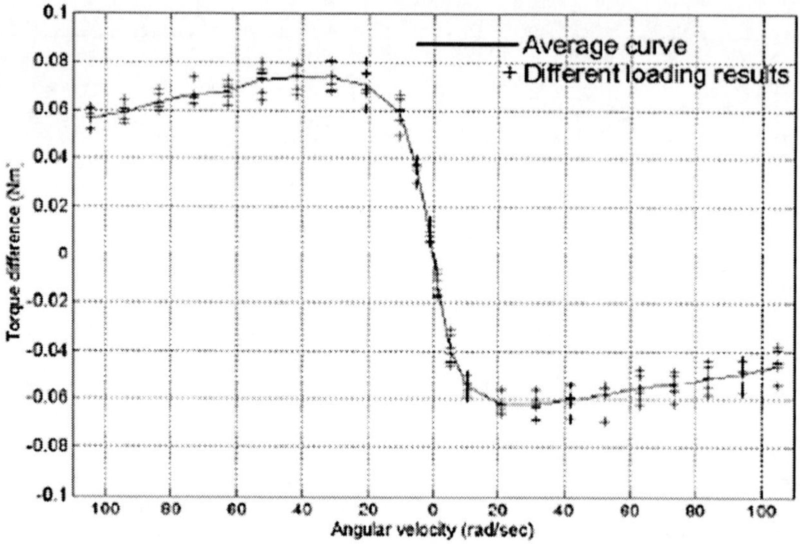

Figure 3.11. Average torque difference values under different loading conditions.

Table 3.2. Model parameters for X-axis feed drive servomechanism

Parameters	Positive direction	Parameters	Negative direction
a_1^{pos}	-1.1115×10^{-2}	a_1^{neg}	-1.0157×10^{-2}
a_2^{pos}	7.7909×10^{-4}	a_2^{neg}	-6.0354×10^{-4}
a_3^{pos}	-2.8408×10^{-5}	a_3^{neg}	-2.0033×10^{-5}
a_4^{pos}	5.8309×10^{-7}	a_4^{neg}	-3.9029×10^{-7}
a_5^{pos}	-6.7334×10^{-9}	a_5^{neg}	-4.3765×10^{-9}
a_6^{pos}	4.0725×10^{-11}	a_6^{neg}	-2.6058×10^{-11}
a_7^{pos}	-1.0019×10^{-13}	a_7^{neg}	-6.3669×10^{-14}
Estimation error ε	3.7764×10^{-3}	Estimation error ε	3.2321×10^{-3}

Similarly, the velocity-dependent friction model for the Y-axis feed drive servomechanism in the experimental setup was obtained as shown in Figure 3.12 and Table 3.3. In this chapter, the velocity-dependent friction models shown in Table 3.2 and Table 3.3 were used to compensate for the velocity-dependent frictions of the X-axis and Y-axis feed drive servomechanisms.

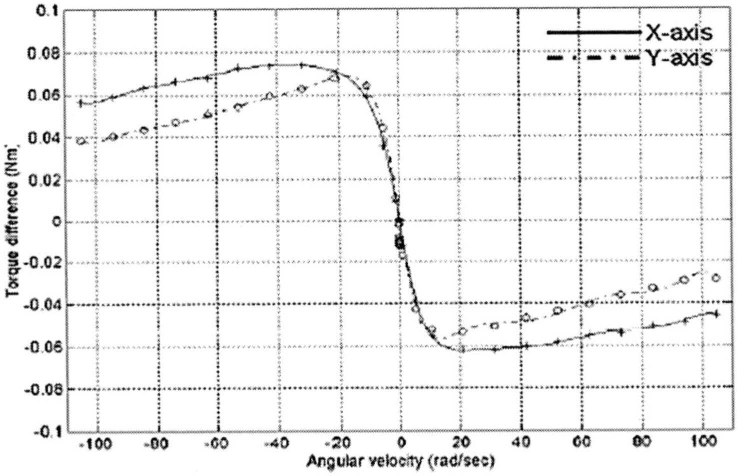

Figure 3.12. Modeling results for velocity-dependent frictions.

Table 3.3. Model parameters for Y-axis feed drive servomechanism

Parameters	Positive direction	Parameters	Negative direction
a_1^{pos}	-1.303×10^{-2}	a_1^{neg}	-1.2636×10^{-2}
a_2^{pos}	1.0904×10^{-3}	a_2^{neg}	-8.7023×10^{-4}
a_3^{pos}	-4.4032×10^{-5}	a_3^{neg}	-3.0113×10^{-5}
a_4^{pos}	9.6360×10^{-7}	a_4^{neg}	-5.8771×10^{-7}
a_5^{pos}	-1.1618×10^{-8}	a_5^{neg}	-6.5091×10^{-9}
a_6^{pos}	7.2499×10^{-11}	a_6^{neg}	-3.8086×10^{-11}
a_7^{pos}	-1.8276×10^{-13}	a_7^{neg}	-9.1316×10^{-14}
Estimation error ε	8.0969×10^{-3}	Estimation error ε	1.6037×10^{-3}

The averaged values of the torque differences, $(\tau_{mj}^{pos}-\tau_{mb}^{pos})$ and $(\tau_{mj}^{neg}-\tau_{mb}^{neg})$ for $j=1,\ldots,n$, were used to compute the model parameters, as shown in Eq. (3.21). Figure 3.13 shows the modeling results and Table 3.4 lists the obtained parameters, $(\alpha_0^{pos}, \alpha_1^{pos}, \alpha_2^{pos}, \omega_0^{pos})$ and $(\alpha_0^{neg}, \alpha_1^{neg}, \alpha_2^{neg}, \omega_0^{neg})$.

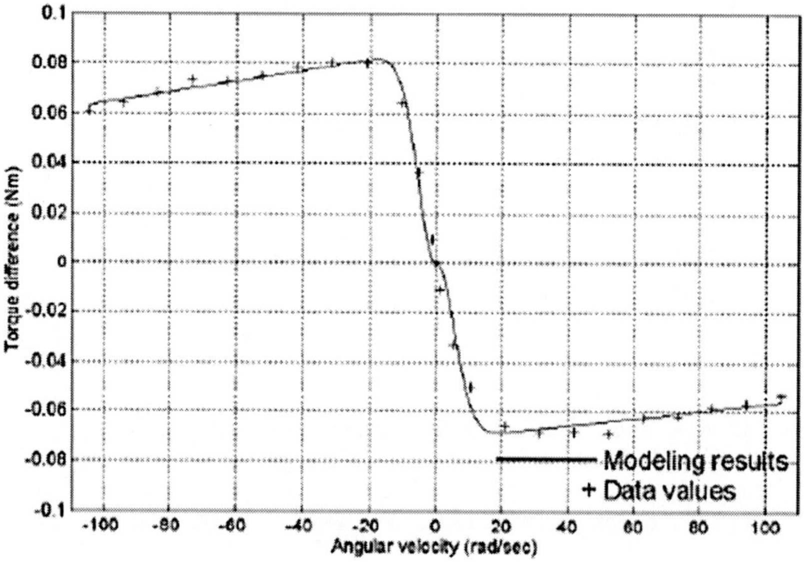

Figure 3.13. Modeling results for velocity-dependent frictions.

Table 3.4. Model parameters obtained by proposed approaches

Parameters in positive direction		Parameters in negative direction	
α_0^{pos}	$\tau_e^{pos} - \alpha_1^{pos}$ τ_e^{pos}: Figure 3.3(a)	α_0^{neg}	$-\tau_e^{neg} - \alpha_1^{neg}$ τ_e^{neg}: Figure 3.3(b)
α_1^{pos}	7.1662×10^{-2}	α_1^{neg}	8.5883×10^{-2}
α_2^{pos}	1.4802×10^{-4}	α_2^{neg}	2.1432×10^{-4}
ω_0^{pos}	7.93	ω_0^{neg}	7.65

3.3. Integration of Obtained Position-Dependent and Velocity-Dependent Frictions

The experimental results presented in the previous two subsections confirm that the friction in a feed drive servomechanism is both position-dependent and velocity-dependent. Thus, it is necessary to develop a friction model that describes the friction as a function of both the position and velocity. Further, because both the position-dependent friction and velocity-dependent friction are clearly repeatable, they can be compensated by considering the average values of the frictions obtained during the experiment. The experimental results also show the asymmetric property of the friction. Thus, distinct functions for different directions are considered to overcome this asymmetric property. Therefore, in this chapter, a friction model to compensate for the friction of the X-axis feed drive servomechanism is developed as

$$\tau_f^{\mathrm{model}} = \begin{cases} \tau_f^{neg}(\theta) + \sum_{i=1}^{m\_neg} a_i^{neg}\omega^i, & \omega \le -\omega_{sb}^{neg} \\ \tau_f^{neg}(\theta), & -\omega_{sb}^{neg} < \omega < 0 \\ 0, & \omega = 0 \\ \tau_f^{pos}(\theta), & 0 < \omega < \omega_{sb}^{pos} \\ \tau_f^{pos}(\theta) + \sum_{i=1}^{m\_pos} a_i^{pos}\omega^i, & \omega \ge \omega_{sb}^{pos} \end{cases} \tag{3.23}$$

where τ_f^{model} denotes the estimated friction. θ and ω are the motor shaft angular position and angular velocity, respectively. As shown in Eq. (3.17), ω_{sb}^{neg} and ω_{sb}^{pos} are the base velocity responses in the negative and positive motion directions, respectively. The coefficients, $\left[a_1^{neg} \ a_2^{neg} \ \cdots \ a_{m\_neg}^{neg}\right]^T$ and $\left[a_1^{pos} \ a_2^{pos} \ \cdots \ a_{m\_pos}^{pos}\right]^T$, are obtained as shown in Table 3.2. The position-dependent friction, $\tau_f^{pos}(\theta)$, is obtained from Figure 3.5(a) for a given measured position, θ, when velocity ω is positive. The position-dependent friction, $\tau_f^{neg}(\theta)$, is obtained from Figure 3.5(b) for a given measured position, θ, when velocity ω is negative. Similarly, by applying the coefficients shown in Table 3.3 and the friction compensation lookup tables shown in Figure 3.5, a friction model for the Y-axis feed drive servomechanism can be obtained and used for friction compensation.

Because $(\alpha_0+\alpha_1)$ results in static friction τ_e, parameter α_0 can be obtained as

$$\alpha_0=\tau_e-\alpha_1 \tag{3.24}$$

where the static friction, τ_e, can be obtained from the friction compensation lookup tables, as shown in Figure 3.3, for a given angular position, θ. Therefore, by considering the asymmetric effect of friction [12], the velocity-dependent friction model is obtained as

$$\tau_f^{\text{model}}=\begin{cases}\begin{cases}\alpha_0^{neg}=-\tau_e^{neg}-\alpha_1^{neg}\\ (\alpha_0^{neg}+\alpha_1^{neg}e^{-(\frac{\omega}{\omega_0^{neg}})^2})\operatorname{sgn}(\omega)+\alpha_2^{neg}\omega, \omega\le-\omega_{sb}^{neg}\end{cases}\\ \tau_e^{neg} \quad ,-\omega_{sb}^{neg}<\omega<0\\ \tau_e^{pos} \quad ,0<\omega<\omega_{sb}^{pos}\\ \begin{cases}\alpha_0^{pos}=\tau_e^{pos}-\alpha_1^{pos}\\ (\alpha_0^{pos}+\alpha_1^{pos}e^{-(\frac{\omega}{\omega_0^{pos}})^2})\operatorname{sgn}(\omega)+\alpha_2^{pos}\omega\ ,\omega\ge\omega_{sb}^{pos}\end{cases}\end{cases} \tag{3.25}$$

where τ_e^{neg} and τ_e^{pos} are the static frictions for the negative and positive directions, respectively. ω_{sb}^{neg} and ω_{sb}^{pos} are the base velocity responses for the negative and positive motion directions, respectively. The coefficients, $(\alpha_0^{neg},\alpha_1^{neg},\alpha_2^{neg},\omega_0^{neg})$, represent the model parameters in the negative motion direction, and these are usually different from the coefficients that represent the model parameters in the positive motion direction, $(\alpha_0^{pos},\alpha_1^{pos},\alpha_2^{pos},\omega_0^{pos})$.

4. DESIGN OF VELOCITY-BASED FRICTION COMPENSATOR

In general, the feedforward friction compensation structure [1-2] shown in Figure 4.1 is used to control the velocity and position motions of the feed drive servomechanisms in the experimental setup. Here, the torque value for friction compensation, τ_{cmp}, is computed by referring to the velocity command, ω_{cmd}, and a friction model obtained using different modeling methods. However, in some cases, the friction torque compensation structure cannot be directly applied to control feed drive servomechanisms that are actuated by velocity-controlled servomotor packs with built-in velocity control loops, as shown in Figure 4.2(a). Because the velocity-controlled servomotor packs only access velocity input commands, ω_{cmd}, a velocity-based friction compensator, as shown in Figure 4.2(b), is required to reduce the adverse motions induced by frictions. Here, the velocity value for friction compensation, ω_{cmp}, is obtained using a prefilter with input signals computed by referring to the position command, θ_{cmd}, the velocity command, ω_{cmd}, and the friction model obtained in this chapter.

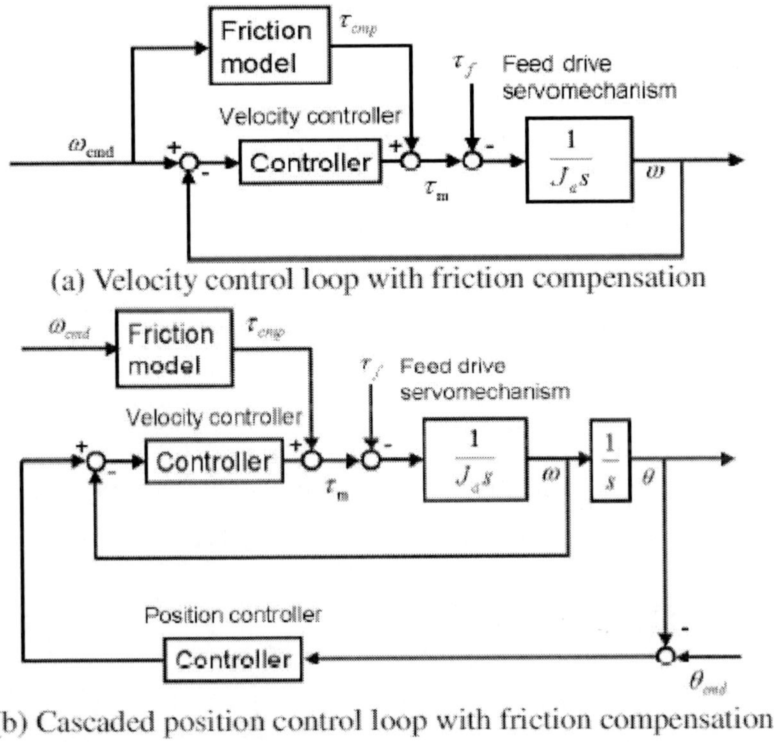

Figure 4.1. Feedforward friction compensation structures.

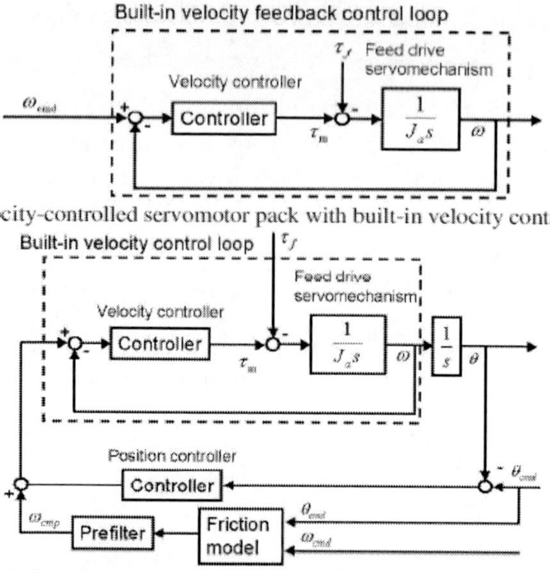

(a) Velocity-controlled servomotor pack with built-in velocity control loop

(b) Cascaded position control loop with velocity-based friction compensator

Figure 4.2. Feedforward friction compensation for velocity-controlled servomotor packs.

Consider the velocity-based friction compensation structure shown in Figure 4.3. Here, $K(s)$ denotes the transfer function of the velocity controller and $G(s)$ denotes the transfer function of the feed drive servomechanism. The velocity command for the built-in velocity control loop, ω_d, is composed of the reference command, ω_r, and the value for friction compensation, ω_{cmp}, as

$$\omega_d = \omega_r + \omega_{cmp} \tag{4.1}$$

Because the velocity output, ω, can be obtained as

$$\omega = \frac{G(s)K(s)}{1+G(s)K(s)}\omega_d - \frac{G(s)}{1+G(s)K(s)}\tau_f, \tag{4.2}$$

by substituting Eq. (4.1) into Eq. (4.2), the velocity output, ω, can be further rewritten as

$$\begin{aligned}\omega &= \frac{G(s)K(s)}{1+G(s)K(s)}\left[\omega_r + \omega_{cmp}\right] - \frac{G(s)}{1+G(s)K(s)}\tau_f \\ &= \frac{G(s)K(s)}{1+G(s)K(s)}\omega_r + \frac{G(s)}{1+G(s)K(s)}\left[K(s)\omega_{cmp} - \tau_f\right]\end{aligned} \tag{4.3}$$

Therefore, in order to reduce the effects induced by friction, τ_f, the value for friction compensation, ω_{cmp}, can be designed as

$$\omega_{cmp} = K(s)^{-1}\tau_f \tag{4.4}$$

In this chapter, because friction τ_f is estimated by applying the modeling and estimation methods mentioned in the previous sections, Eq. (4.4) is rewritten as

$$\omega_{cmp} = K(s)^{-1}\tau_f^{\text{model}} \tag{4.5}$$

where the estimated friction, τ_f^{model}, is obtained by Eq. (3.23) and filter $K(s)^{-1}$ denotes the prefilter, as shown in Figure 4.2(b).

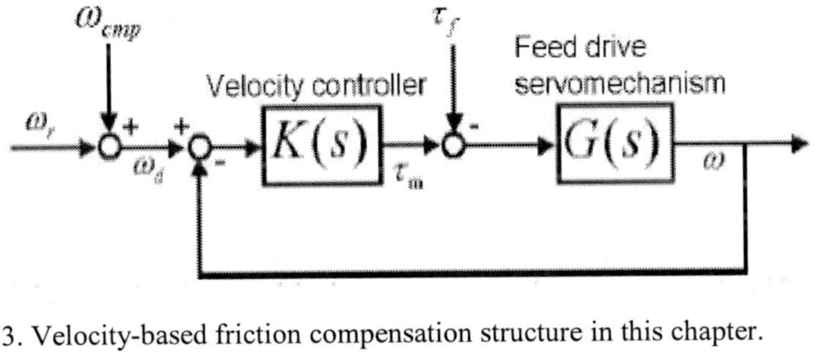

Figure 4.3. Velocity-based friction compensation structure in this chapter.

In a case where the velocity controller is implemented as a PI controller, the transfer function, $K(s)$, is obtained as

$$K(s) = K_p + K_i\frac{1}{s} \tag{4.6}$$

where K_p denotes the proportional gain and K_i denotes the integral gain of the PI controller. Therefore, the prefilter, $K(s)^{-1}$, can be obtained as

$$K(s)^{-1} = \frac{s}{K_p s + K_i} \tag{4.7}$$

Eq. (4.7) also represents the transfer function of a low-pass-filtered derivative operator with bandwidth $\frac{K_i}{K_p}$. However, for most industrial applications, the design of the velocity controller in the built-in velocity control loop is usually unknown in advance, and thus it is difficult to design prefilter $K(s)^{-1}$ systematically. Therefore, in this chapter, a PI controller is used to approximate the velocity controller in a built-in velocity control loop. The corresponding gains, K_p and K_i, are then obtained using several identification methods, and thus the prefilter design shown in Eq. (4.7) becomes realizable. Further, the gains, K_p and K_i, are fine-tuned manually in order to minimize the adverse effects induced by frictions. Figure 4.4 shows the structure for identifying the control gains, K_p and K_i. For the given estimated friction, τ_f^{model}, and the equivalent inertia, $\hat{J}_a$, an auto-tuning process is applied to tune the control gains, K_p and K_i, in order to minimize the difference between the actual output, ω_{actual}, and model output, ω_{model}. Here, the estimated friction, τ_f^{model}, is obtained by Eq. (3.23), and the equivalent inertia, $\hat{J}_a$, can be obtained by considering the relations between the driving torque and acceleration output of the tested feed drive servomechanism.

Figure 4.5 shows the identification results for the equivalent inertia, $\hat{J}_a$, under a no-load condition. Equivalent inertia values of $0.57 \times 10^{-4}\ \text{kg} \cdot \text{m}^2$ and $0.59 \times 10^{-4}\ \text{kg} \cdot \text{m}^2$ are obtained for the X-axis and Y-axis feed drive servomechanisms of the experimental setup, respectively. By applying the estimated friction, τ_f^{model}, and the equivalent inertia, $\hat{J}_a$, control gains (K_p, K_i) of $(0.012, 0.26)$ and $(0.01, 0.34)$ are obtained for the X-axis and Y-axis feed drive servomechanisms of the experimental setup, respectively. Figure 4.6 shows the identification results.

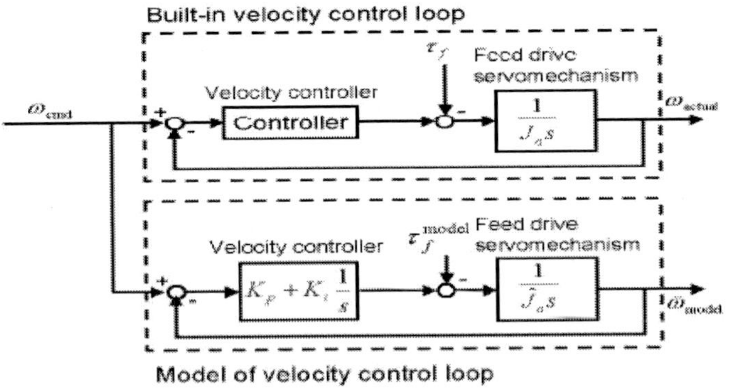

Figure 4.4. Structure for identification of control gains K_p and K_i.

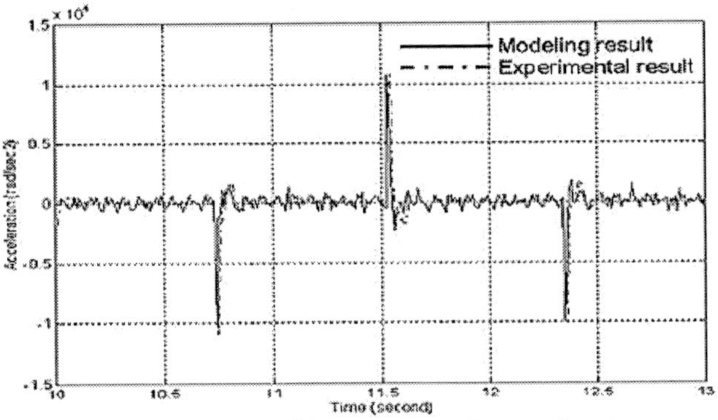

(a) X-axis ($\hat{J}_a = 0.57 \times 10^{-4}$ kg · m^2)

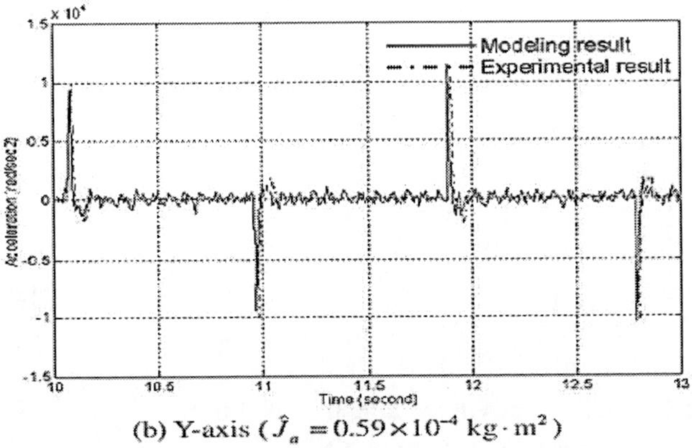

(b) Y-axis ($\hat{J}_a = 0.59 \times 10^{-4}$ kg · m^2)

Figure 4.5. Identification results for equivalent inertia under no-load condition.

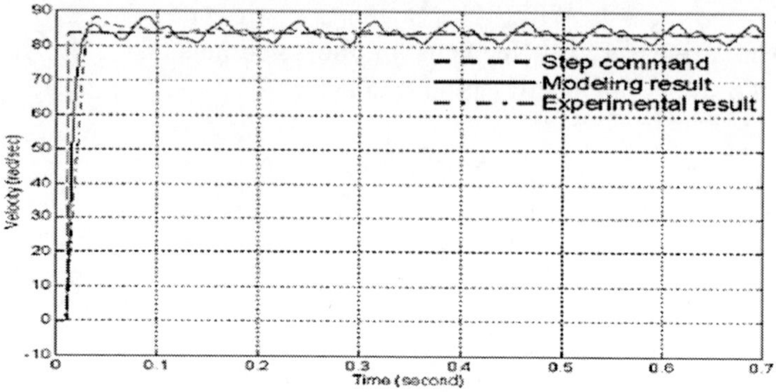

(a) X-axis ($K_p = 0.012$ and $K_i = 0.26$)

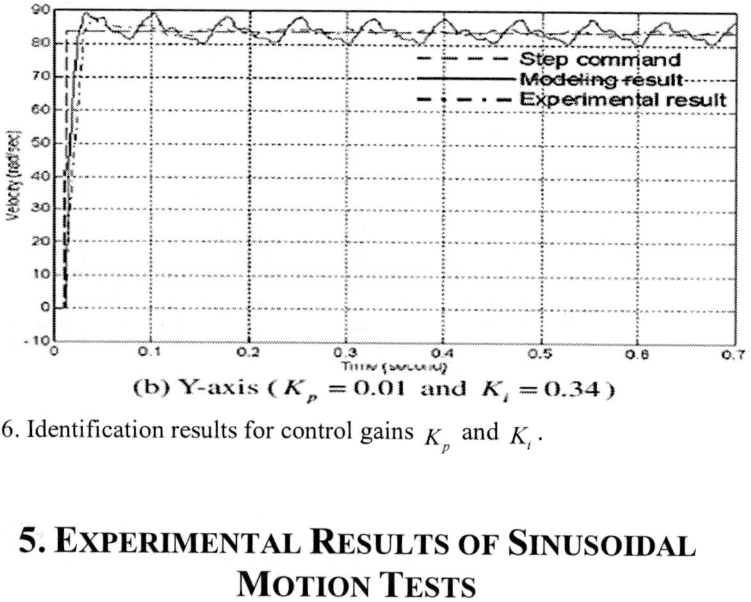

(b) Y-axis ($K_p = 0.01$ and $K_i = 0.34$)

Figure 4.6. Identification results for control gains K_p and K_i.

5. Experimental Results of Sinusoidal Motion Tests

In this chapter, sinusoidal motion tests were carried out on the experimental setup in order to show the feasibility of the proposed friction identification method. Sinusoidal velocity commands with an amplitude 52.36 rad/sec and frequencies of 0.5 Hz and 1.0 Hz were used for the sinusoidal motion tests. The tracking errors are considered to evaluate the motion accuracy. Moreover, to evaluate the different friction models obtained by the conventional method and the method proposed in this chapter, the feedforward friction compensation structure shown in Figure 5.1 was used to control the motions of the feed drive servomechanisms in the experimental setup. Here, the torque value for friction compensation, τ_{cmp}, is computed by referring to the velocity command, ω_{cmd}, and the different friction models obtained using different identification methods.

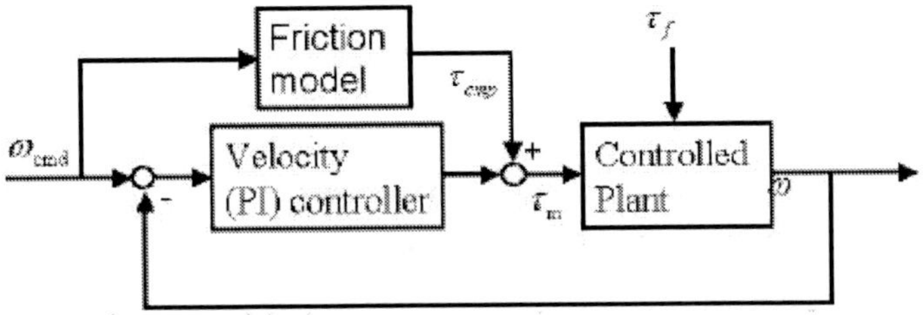

Figure 5.1. Feedforward friction compensation and velocity feedback control.

Figure 5.2 shows the friction-velocity map obtained by conventional approaches and Table 5.1 lists the corresponding model parameters. Here, the friction model shown in Eq. (3.18) is adopted to compare the different identification methods. Parameters α_1^{pos} and α_1^{neg} are further modified in order to consider the static frictions shown in Figure 3.3. The values in brackets are the average values of α_0.

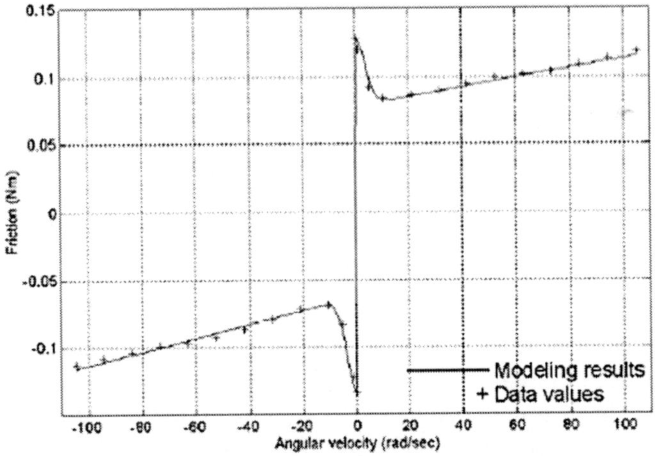

Figure 5.2. Obtained friction-velocity map and friction models.

In this chapter, the sinusoidal motions obtained by applying different friction models are denoted as follows:

- UNCOMP: denotes the motion results obtained without friction compensation.
- COMP_C: denotes the motion results obtained using the friction model listed in Table 5.1.
- COMP_P: denotes the motion results obtained using the friction model listed in Table 3.4.

Table 5.1. Friction model parameters obtained by conventional approaches

Parameters in positive direction		Parameters in negative direction	
α_0^{pos}	$\tau_e^{pos} - \alpha_1^{pos}$ (7.9×10^{-2})	α_0^{neg}	$-\tau_e^{neg} - \alpha_1^{neg}$ (6.3×10^{-2})
α_1^{pos}	4.9660×10^{-2}	α_1^{neg}	7.0987×10^{-2}
α_2^{pos}	3.4532×10^{-4}	α_2^{neg}	5.1030×10^{-4}
ω_0^{pos}	5.2	ω_0^{neg}	5.1

Here, ten iterations were used for each motion test to obtain average motion performances. Several performance indices were used to evaluate the steady motion performances as follows:

- RMS: denotes the root-mean-square value of errors.
- MAX: denotes the absolute value of maximum errors.
- STD: denotes the standard deviation of data values.

Figure 5.3 shows the experimental results of different sinusoidal motion tests and Table 5.2 lists the statistical results of the obtained tracking errors. It is quite obvious that friction significantly deteriorates the motion accuracy when the velocity of the motion is near zero. The MAX values are both over 10.0 rad/sec for the motion tests. However, by applying friction compensation, the MAX values are reduced, with reduction rates of 24.9% and 36.94% achieved for the motion tests with frequencies of 0.5 Hz and 1.0 Hz, respectively. In addition, RMS values of 26.97% and 44.16% are achieved for the different motion tests. Here, because the motion test with a frequency of 0.5 Hz contains more low-velocity motions that could be significantly affected by friction, the following two facts are observed:

1) The STD values for the motion test with a frequency of 0.5 Hz are generally larger than those for the motion test with a frequency of 1.0 Hz.
2) The rate of reduction for the motion test with a frequency of 0.5 Hz is generally less than that for the motion test with a frequency of 1.0 Hz.

These observations follow the fact that the friction compensation performance is usually limited for low-velocity motions because of 1) the difficulty in estimating friction characteristics at a low velocity and 2) the validity of the selected friction model at a low velocity. However, the experimental results show that friction compensation can significantly improve the motion accuracy of feed drive servomechanism motions. In comparison with conventional identification methods, although the proposed identification method slightly increases the maximum tracking errors, the RMS values are reduced, with a reduction rate of 9.52% achieved in the experiments.

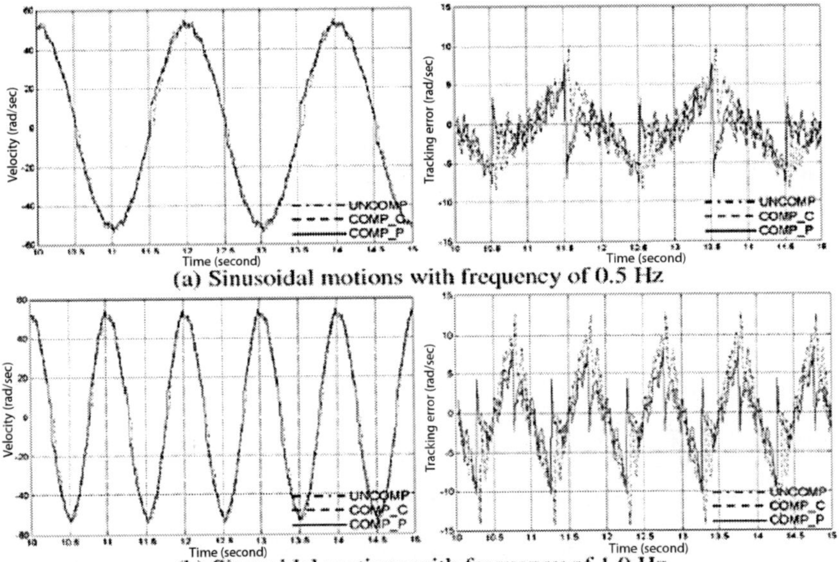

(a) Sinusoidal motions with frequency of 0.5 Hz

(b) Sinusoidal motions with frequency of 1.0 Hz

Figure 5.3. Experimental results of sinusoidal motion tests.

Here, the STD values for the motion test with a frequency of 0.5 Hz are less than 1.5% of the corresponding average values, and the STD values for the motion test with a frequency of 1.0 Hz are less than 1.0% of the corresponding average values. Thus, the experimental results show the feasibility of the friction-modeling method proposed in this chapter.

Table 5.2. Statistical results of tracking errors for different motion tests

	0.5 Hz		1.0 Hz	
	RMS (rad/sec) STD (%)	MAX (rad/sec) STD (%)	RMS (rad/sec) STD (%)	MAX (rad/sec) STD (%)
UNCOMP	3.6264 0.18	10.0807 0.26	6.3411 0.21	12.7780 0.53
COMP_C	2.9273 1.41	7.5704 1.27	3.7767 0.12	8.0580 0.58
COMP_P	2.6485 1.43	7.6454 1.25	3.5410 0.16	8.4200 0.63

6. Experimental Results of Circular Motion Tests

Because circular motions are usually employed to evaluate the motion performances of a CNC milling machine, circular motion tests were carried out on the experimental setup to show the feasibility of the friction modeling and compensation methods proposed in this chapter. Circular motion commands with a radius of 8 mm and feedrates of 100.0 mm/min and 500.0 mm/min were used to conduct the circular motion tests. The contouring errors are considered to evaluate the motion accuracy. Different loading conditions, with no load and a 5.0-kg weight, are also considered in this chapter. For the sake of comparison, two motion results are denoted as follows:

- NoComp: denotes the motion results obtained without using friction compensation.
- VeComp: denotes the motion results obtained using the modeling and compensation methods developed in this chapter.

Moreover, three performance indices are used to evaluate the steady motion performances as follows:

- AVG: denotes the average value of errors.
- RMS: denotes the root-mean-square value of errors.
- MAX: denotes the maximum value of absolute errors.

In order to minimize the adverse effects induced by frictions, the control gains were further fine-tuned manually by considering the MAX value of the contouring errors.

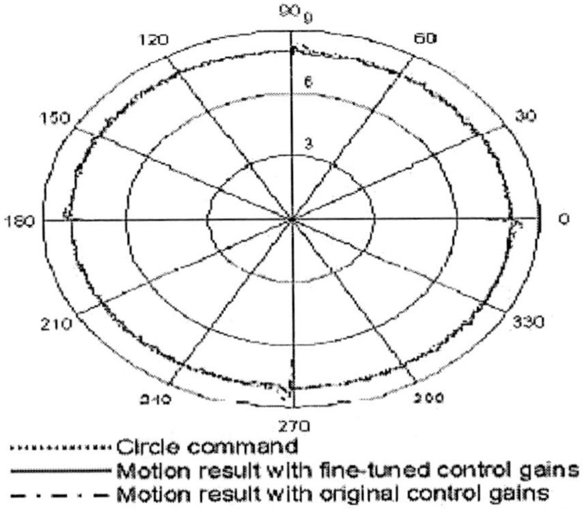

(a) Contouring results (Contouring errors are multiplied by 20)

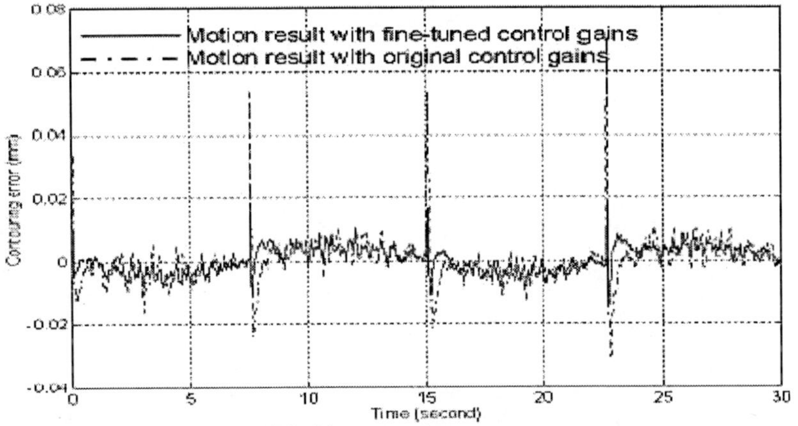

(b) Contouring error

Figure 6.1. Circular motions with different control gains (feedrate 100.0 mm/min, no load).

Control gains (K_p, K_i) of $(0.045, 0.15)$ and $(0.049, 0.12)$ were thus obtained for the X-axis and Y-axis feed drive servomechanisms of the experimental setup, respectively. The MAX value of contouring errors was significantly reduced from 0.0710 mm to 0.0185 mm, which is a reduction rate of 73.94%. Figure 6.1 shows the motion results obtained by applying different control gains.

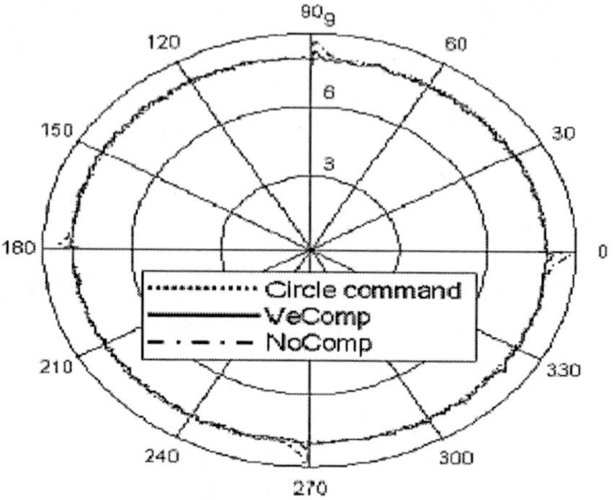

(a) Contouring results (Contouring errors are multiplied by 20)

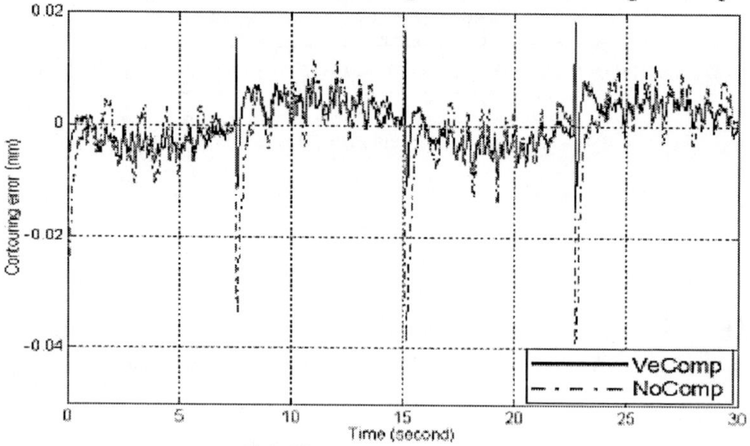

(b) Contouring error

Figure 6.2. Experimental results for circular motion tests with feedrate of 100.0 mm/min under no-load condition.

Figures 6.2–6.5 show the experimental results for the circular motion tests with different feedrate and payload conditions. In the experiments, by applying the velocity-based friction compensation structure developed in this chapter, the circular motions with a slow feedrate achieved the best improvements on the RMS and MAX values of the contouring errors. The RMS value of the contouring errors was reduced from 0.0067 mm to 0.0039 mm, which is a reduction rate of 41.79%. The MAX value of the contouring errors was reduced from 0.0448 mm to 0.0185 mm, which is a reduction rate of 58.71%.

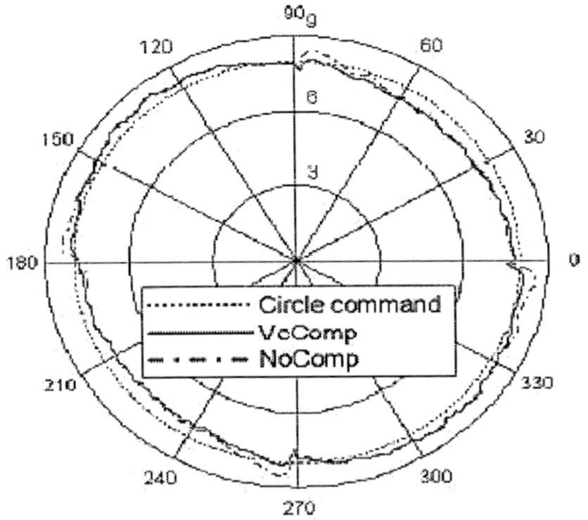

(a) Contouring results (Contouring errors are multiplied by 20)

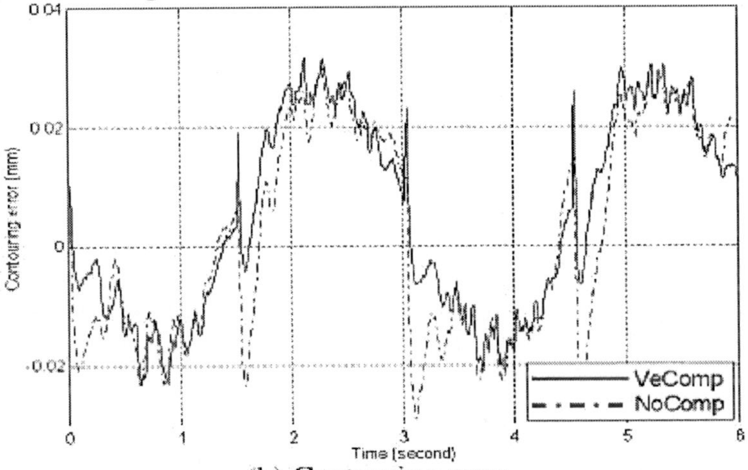

(b) Contouring error

Figure 6.3. Experimental results for circular motion tests with feedrate of 500.0 mm/min under no-load condition.

However, because of the significant servo-lags and mismatched dynamics that usually exist in the rapid motions of feed drive servomechanisms, the improvement in contouring accuracy became ambiguous when the motion tests were performed using a fast feedrate. Nevertheless, as shown in Figures

6.2–6.5, the velocity-based friction compensation structure could successfully reduce the magnitude of quadrant glitches during circular motions. It is well known that the contouring accuracy can decrease when adding payload to a feed drive servomechanism. However, the velocity-based friction compensation structure could also reduce the contouring errors during motion tests. The RMS value of the contouring errors was reduced from 0.0072 mm to 0.0046 mm, which is a reduction rate of 36.11%. The MAX value of the contouring errors was reduced from 0.0464 mm to 0.0192 mm, which is a reduction rate of 58.63%.

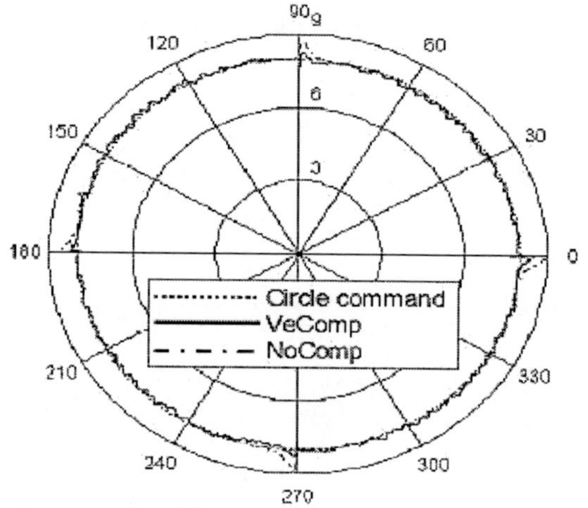

(a) Contouring results (Contouring errors are multiplied by 20)

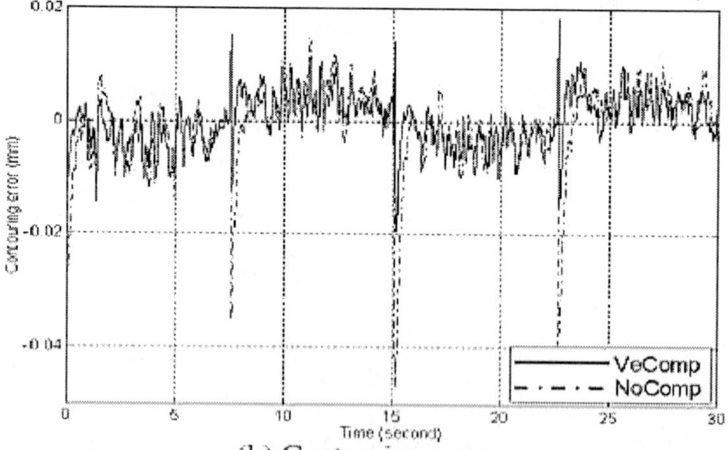

(b) Contouring error

Figure 6.4. Experimental results for circular motion tests with feedrate of 100.0 mm/min under 5.0-kg payload condition.

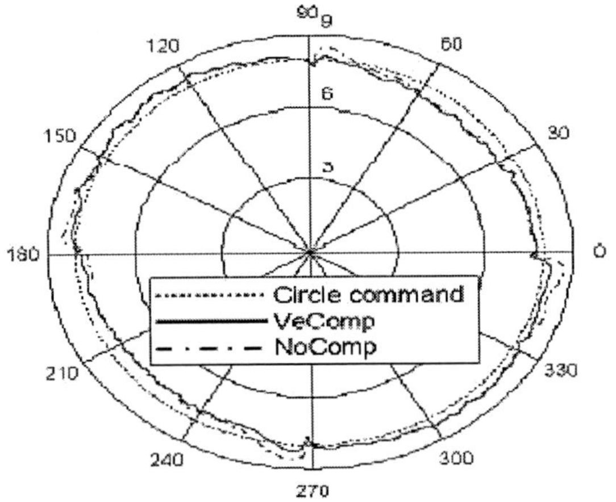

(a) Contouring results (Contouring errors are multiplied by 20)

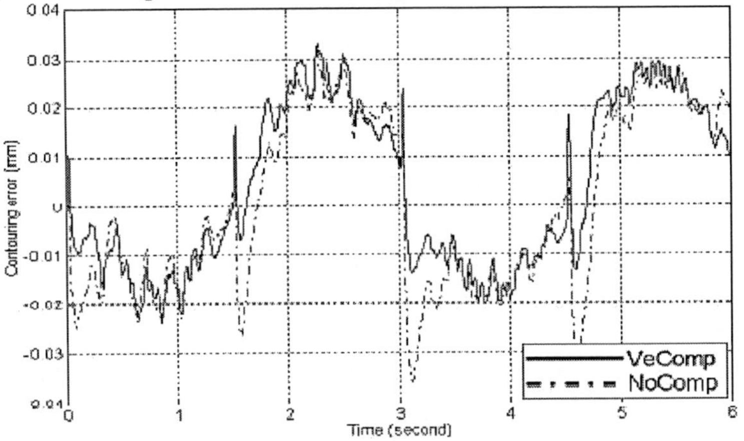

(b) Contouring error

Figure 6.5. Experimental results for circular motion tests with feedrate of 500.0 mm/min under 5.0-kg payload condition.

CONCLUSION

Frictions are physical phenomena that exist in the feed drive servomechanisms of CNC machine tools and usually cause adverse effects to the motions of the feed drive servomechanisms. To reduce these adverse effects, several model-based friction compensation methods such as feedforward compensation methods and nonlinear feedback compensation methods [32] are usually considered in the motion control design of mechanical systems. Therefore, good friction models are usually required, and compensation performances heavily depend on the accuracy of these applied friction models. Although many experiments such as the breakaway experiment and constant-velocity experiment could be used to identify the friction characteristics of the applied feed drive servomechanism, some problems still exist and could significantly affect the friction identification results. For instance, in practical applications, there are several contact surfaces in a feed drive servomechanism, which could generate frictions with different characteristics. Some considerable position-dependent perturbations, usually caused by the surface irregularities, assembly errors, and geometric machining errors of the mechanical parts, exist in a feed drive servomechanism and could deteriorate the obtained friction-velocity map used for friction estimation. The unmodeled dynamics, usually induced by several flexible and inertial elements in feed drives, increase the uncertainties of the estimated friction characteristics. However, those important factors that could significantly affect the friction identification results are usually ignored in friction studies for the motion control systems of CNC machine tools.

This chapter aimed at developing a new friction identification method that considers the adverse effects induced by the position-dependent perturbations and unmodeled dynamics of a feed drive servomechanism in a CNC machine tool. In this chapter, a modified velocity control structure comprised of a PI feedback control and DOB was developed to compensate for the speed perturbations induced by external disturbances and system uncertainties during constant-velocity experiments. The experimental results show that the velocity responses are kept almost constant during steady motions. The variance value of a steady velocity response was significantly reduced from 1.4108×10^{-2} rad/sec to 1.4750×10^{-3} rad/sec. Moreover, using the friction and velocity data values obtained in constant-velocity experiments, a new friction extraction method was developed in this chapter to remove the adverse effects induced by position-dependent perturbations by considering the subtraction of different friction data values. By applying the modified velocity control structure and newly-developed friction extraction method, more reliable friction and velocity data values could be obtained and used to construct a friction model that reduces the uncertainties of the estimated friction characteristics. Several experiments and motion tests were carried out on a three-axis CNC milling machine, and the feedforward friction compensation structure was applied to control the motions of its feed drive servomechanism to evaluate the different friction models obtained using different identification methods. The experimental results of the motion tests indicated that the friction compensated motion control system with the friction model obtained in this chapter achieved reductions of 44.16% (the best case) and 26.97% (the worse case) in the RMS values of the tracking errors. Moreover, in comparison with conventional methods, the RMS values of the tracking errors were further reduced, with a reduction rate of 9.52%. The experimental results thus show the feasibility of using the identification method proposed in this chapter in motion control design applications for CNC machine tools.

In some industrial applications, a feedforward friction torque compensation structure cannot be directly used to reduce the adverse effects induced by frictions. Therefore, a new velocity-based friction compensation structure was developed in this chapter to reduce the friction effects using an additional velocity command obtained by a prefilter with input signals computed by referring to the position command, velocity command, and friction model obtained in this chapter. Moreover, the parameters of the prefilter can be systematically identified and fine-tuned in order to minimize the motion errors induced by frictions. The velocity-based friction compensation structure was applied to control the motions of the feed drive servomechanisms in a CNC milling machine to show the feasibility of the approaches proposed in this chapter. The experimental results of circular motion tests indicated that the friction modeling and compensation methods developed in this chapter achieved a reduction of 41.79% in the RMS value of contouring errors and a reduction of 58.71% in the MAX value of contouring errors during circular motions with slow feedrate commands. The experimental results thus show the feasibility of using the modeling and compensation methods proposed in this chapter in motion control design applications for CNC machine tools.

In this chapter, the static friction and static maps between the velocity and frictions were used to describe the friction characteristics of a feed drive servomechanism. However, many studies [3-5,12] have reported that frictions can exhibit properties such as the Dahl effect [33], friction lag [12], pre-sliding displacement [18], and stick-slip motion [34] that cannot be described by static friction models because of the internal dynamics of frictions. Therefore, although the experimental results showed an improvement in circular motions from using the static friction modeling and compensation methods proposed in this chapter, the motion performances could be further improved by considering the dynamic properties of frictions.

REFERENCES

[1] Tsai, M. C., Chiu, I. F., and Cheng, M. Y. (2004). Design and implementation of command and friction feedforward control for CNC motion controllers. *IEE Proceedings: Control Theory and Applications, 151 1*, 13-20.

[2] Elfizy, A. T., Bone, G. M., and Elbestawi, M. A. (2004). Model-based controller design for machine tool direct feed drives. *International Journal of Machine Tools and Manufacture, 44 5*, 465-477.

[3] Shih, Y. T. and Lee, A. C. (2003). Survey on modeling and control for motion systems with friction. *Journal of the Chinese Society of Mechanical Engineers, Transactions of the Chinese Institute of Engineers, Series C, 24 4*, 337-352.

[4] Armstrong-Helouvry, B., Dupont, P., and Canudas de Wit, C. (1994). A survey of models, analysis tools and compensation methods for the control of machines with friction. *Automatica, 30 7*, 1083-1138.

[5] Awrejcewicz, J. and Olejnik, P. (2005). Analysis of dynamic systems with various friction laws. *Applied Mechanics Reviews, 58 1-6*, 389-410.

[6] Chen, C. L., Jang, M. J., and Lin, K. C. (2004). Modeling and high-precision control of a ball-screw-driven stage. *Precision Engineering, 28 4*, 483-495.

[7] Persianoff, R., Ray, P., and Vidal, O. (2003). Comparison between an experimental study and a numerical model of the dynamic behaviour of machine-tool slideways. *Proceedings of the Institution of Mechanical Engineers, Part B: Journal of Engineering Manufacture, 217 8*, 1111-1115.

[8] Hsieh, C. and Pan, Y. C. (2000). Dynamic behavior and modelling of the pre-sliding static friction. *Wear, 242 1-2*, 1-17.

[9] Polycarpou, A. A. and Soom, A. (1996). A two-component mixed friction model for a lubricated line contact. *Journal of Tribology, Transactions of the ASME, 118*, 183-189.

[10] Cetinkunt, S., Yu, W. L., Filliben, J., and Donmez, A. (1994). Friction characterization experiments on a single point diamond turning machine tool. *International Journal of Machine Tools and Manufacture, 34 1*, 19-32.

[11] Marui, E. and Endo, H. (1992). Significance of contact resistance in boundary lubrication. *Wear, 156 1*, 49-55,.

[12] Armstrong-Helouvry, B. (1991). *Control of machines with friction.* Boston: Kluwer.

[13] Armstrong, B. S. R. and Chen, Q. (2008). The Z-properties chart. *IEEE Control Systems Magazine*, *28 5*, 79-89.

[14] Cheok, K. C., Hu, H., and Loh, N. K. (1988). Modeling and identification of a class of servomechanism systems with stick-slip friction. *Journal of Dynamic Systems, Measurement and Control, Transactions of the ASME*, *110 3*, 324-328.

[15] Karnopp, D. (1985). Computer simulation of stick-slip friction in mechanical dynamic systems. *Journal of Dynamic Systems, Measurement and Control, Transactions of the ASME*, *107 1*, 100-103.

[16] Kim, J. H., Chae, H. K., Jeon, J. Y., and Lee, S. W. (1996). Identification and control of systems with friction using accelerated evolutionary programming. *IEEE Control Systems Magazine*, *16 4*, 38-47.

[17] Canudas de Wit, C. and Lischinsky, P. (1997). Adaptive friction compensation with partially known dynamic friction model. *International Journal of Adaptive Control and Signal Processing*, *11 1*, 65-80.

[18] Canudas de Wit, C., Olsson, H., Astrom, K. J., and Lischinsky, P. (1995). A new model for control of systems with friction. *IEEE Transactions on Automatic Control*, *40 3*, 419-425.

[19] Popovic, M. R. and Goldenberg, A. A. (1998). Modeling of friction using spectral analysis. *IEEE Transactions on Robotics and Automation*, *14 1*, 114-122.

[20] Ohnishi, K. (1987). A new servo method in mechatronics. *Transactions of the Japan Society for Electrical Engineering*, *107 D*, 83-86.

[21] Kobayashi, S., Awaya, I., Kuromaru, H., and Oshitani, K. (1995). Dynamic model based auto-tuning digital servo driver. *IEEE Transactions on Industrial Electronics*, *42*, 462-466.

[22] *Panasonic MINAS A4-series instruction manual* (DV0P4210) (2004). Matsushita Electric Industrial Co., Ltd.

[23] Söderström, T., and Stoica, P. (1989). *System Identification*. New York: Prentice Hall.

[24] Umeno, T. and Hori, Y. (1991). Robust speed control of servomotors using modern two degrees-of-freedom controller design. *IEEE Transactions on Industrial Electronics*, *38*, 363-368.

[25] Lee, H. S. and Tomizuka, M. (1996). Robust motion controller design for high accuracy positioning systems. *IEEE Transactions on Industrial Electronics*, *43*, 48-55.

[26] Choi, B. K., Choi, C. H., and Lim, K. (1999). Model-based disturbance attenuation for CNC machining centers in cutting process. *IEEE/ASME Transactions on Mechatronics*, *4 2*, 157-168.

[27] Canudas de Wit, C., Noel, P., Aubin, A., and Brogliato, B. (1991). Adaptive friction compensation in robot manipulators: Low velocities. *International Journal of Robotics Research*, *10 3*, 189-199.

[28] Oppenheim, A. V., and Schafer, R. W. (1989). *Discrete-time signal processing*. New Jersey: Prentice Hall.

[29] Garcia, E., De Santos, P. G., and Canudas de Wit, C. (2002). Velocity dependence in the cyclic friction arising with gears. *International Journal of Robotics Research*, *21 9*, 761-771.

[30] Lischinsky, P., Canudas de Wit, C., and Morel, G. (1999). Friction compensation for an industrial hydraulic robot. *IEEE Control Systems Magazine*, *19 1*, 25-32.

[31] Grace, A. (1994). *Optimization toolbox for use with MATLAB, User's Guide*. The MathWorks Inc.

[32] Canudas de Wit, C. and Kelly, R. (2007). Passivity analysis of a motion control for robot manipulators with dynamic friction. *Asian Journal of Control*, *9 1*, 30-36.

[33] Dahl, P. R. (1976). Solid friction damping of mechanical vibrations. *AIAA Journal*, *14 12*, 1675-1682.

[34] Kato, S., Yamaguchi, K., and Matsubayashi, T. (1974). Stick-slip motion of machine tool slideway. *Journal of Engineering for Industry, Transactions of the ASME*, *96 2*, 557-566.

In: Industrial Control Systems
Editors: R. Gilbert and A. Schultz
ISBN: 978-1-61209-988-0

Chapter 2

DYNAMIC MATRIX CONTROL WITH INTERNAL MODEL BASED ON ANN OF A CONTINUOUS EXTRACTIVE PROCESS FOR BIOETHANOL PRODUCTION

Edwin G. Boza Condorena[1], Daniel Ibraim Pires Atala[2] and Aline Carvalho da Costa[1]

[1]Laboratory of Optimization, Design and Advanced Control, School of Chemical Engineering, State University of Campinas, 13083-970, Campinas, SP, Brazil

[2]Centro de Tecnologia Canavieira, 13400-970-Piracicaba,SP, Brasil

ABSTRACT

The MPC (Model Predictive Control) control algorithms have been well accepted for industrial applications (Dechechi, 1998; Morari and Lee, 1999; Piché et al., 2000; Tatjewski, 2006). Among the advantages of these algorithms is the fact that they deal directly with interactions between inputs and outputs and are easily applied to a multivariable process (MIMO - Multiple Input Multiple Output) (Santos and Biegler, 1999). In addition, restrictions on inputs and outputs due to physical constraints and security, common in process industries, can be directly integrated into the MPC algorithm. One of the most popular predictive control algorithms is the DMC or dynamic matrix control (Qin and Badgwell, 1997; Gawkowski, 2007), which is based on a linear model.

One difficulty in using linear control algorithms is that the majority of the chemical processes exhibit a nonlinear behavior. Thus, controllers based on linear models must have their parameters adjusted in a conservative manner to ensure stability in a wide range of operating conditions, which can result in a very slow control (Chu et al., 1998). In some cases, even with this conservative adjustment of parameters, the linear controller does not provide satisfactory performance. Some examples where this can occur are non-stationary processes in which an optimal profile is implemented (Gattu and Zafiriou, 1999), systems that exhibit complex interactions with constraints and highly nonlinear phenomena (Santos and Biegler, 1999), and processes subject to large disturbances (Rawlings, 2000).

Another option for improving the performance of the control scheme is the use of predictive control algorithms based on nonlinear models (Le Lann et al., 1999). Several types of models have been used for this purpose, such as differential equations, neural networks, etc. (Morari and Lee, 1999).

Artificial neural networks have been increasingly applied to describe chemical processes due to their ability to approximate highly nonlinear systems involving a large number of state variables. Different structures of neural networks were considered as non-linear models for advanced control algorithms (Fonseca, 1998; Henrique, 1999; Rohani et al., 1999; Kambhampati et al., 2000; Meleiro et al., 2001; Nagy, 2007).

In this chapter, the goal is to study the nonlinear control of a continuous extractive process of bioethanol production. An algorithm based on DMC is proposed, using an internal model based on artificial neural networks to represent the dynamic behavior of the manipulated and controlled variables, in substitution of the linear convolution model of the DMC controller.

The extractive alcoholic fermentation process combines the fermentor and a flash vessel to continuously extract part of the ethanol produced. Experimental data from 400 h of operation of a pilot plant is used to develop the ANN based model for the control system.

NOMENCLATURE

C	Cell viability (%)
D	Dilution rate (h^{-1})
F_0	Feed flow rate to the fermentor (L/h)
F	Fermentor outlet flow rate (L/h)
F_P	Permeate flow rate (L/h)
F_{PU}	Fermentor purge flow rate (L/h)

F_{in}	Flash tank inlet flow rate (L/h)
Frf	Flash tank liquid outlet flow rate (L/h)
F_{re}	Return stream flow rate (L/h)
F_V	Flash tank vapour outlet flow rate (L/h)
G	Fermentor glycerol concentration (g/L)
P	Fermentor ethanol concentration (g/L)
Prf	Ethanol concentration in flash tank return flow rate (g/L)
P_{flash}	Flash tank pressure (mmHg)
T_{flash}	Flash tank temperature (°C)
So	Inlet substrate concentration (g/L)
S	Fermentor substrate concentration (g/L)
Srf	Substrate concentration in flash tank return flow rate (g/L)
T_{ferm}	Fermentor temperature (°C)
t	Time (h)
tr	Residence time (h)
V	Total volume of the system (L)
X	Fermentor biomass concentration ((%v/v)
Xrf	Biomass concentration in flash tank return flow rate (%v/v)

1. INTRODUCTION

The control algorithms MPC (Model Predictive Control) have been well accepted for industrial applications (Dechechi, 1998, Morari and Lee, 1999). According to Piché et al. (2000), these algorithms have been used in multiple applications in the refining, petrochemical, chemical, paper and food industries. Among the advantages of these algorithms is the fact that they deal directly with interaction between inputs and outputs and are easily applied to multivariable processes. In addition, restrictions on inputs and outputs due to physical constraints and security, common in processing industries, can be directly integrated into the MPC algorithm.

In the center of an MPC algorithm is a dynamic model, in the case of the DMC algorithm, a linear convolution model. Until recently, most industrial applications used linear dynamic models. These models were preferred to the non-linear ones due to the difficulty of developing nonlinear models from data of the plant and because of computational requirements involved in using nonlinear models in MPC formulation (Piché et al., 2000). The rapid development in computer power, however, led to the emergence of various control algorithms involving nonlinear models (Costa and Maciel Filho, 2003).

The main problem with the linear controllers is that they are not robust, i.e. they do not maintain performance in different regions of operation or changes in operating conditions (Costa et al., 2002). According to Piché et al. (2000), MPC algorithms based on linear models are acceptable when the process operates at a single setpoint and the use of the controller is mainly for disturbance rejection. However, many chemical processes do not operate in a single setpoint; they are operated under different conditions depending on market requirements. Thus, the control system must be designed to keep the process operating properly at different operating points. Due to the nonlinearity of chemical processes, the changes of the operating condition usually make a linear model which well represents the process in a given operating point to present large errors in another operating point.

One possibility in dealing with this problem is to allow the linear model to adapt to changes in the process, i.e. using adaptive control algorithms (Dechechi, 1998). However, adaptive schemes may present stability problems, especially when frequent disturbances affect the process.

Another option for improving the performance of the control system is the use of predictive control algorithms based on nonlinear models (Le Lann et al., 1999). The neural networks are being increasingly applied to describe chemical processes due to their ability to approximate highly nonlinear systems with a large number of state variables. Hussain (1999) presents a review of the applications of neural networks in control processes. A review of studies in the literature (Fonseca, 1998; Santos et al. (2000); Piché et al. (2000); Kambhampati et al. 2000; Meleiro et al., 2001), shows that the networks are able to describe the system in a large number of cases, and are quite versatile, since they can be incorporated with relative facility in several known strategies for nonlinear control.

One of the most important stages of implementing a nonlinear predictive control algorithm is the model development. When working with neural networks models, there are several difficulties related to their training, e.g. the definition of quality and quantity of training data and the definition of the optimal structure of the network. On the other hand, training algorithms with different convergence criteria can lead to models with different generalization capacities. Finally, the last step is the validation of the model with data not used in training.

Taking this into consideration, in this chapter a DMC control algorithm is used to control an extractive alcoholic fermentation process using a process model based on ANN. The control algorithm was programmed in MATLAB.

1.1. Dynamic Matrix Control (DMC)

The basic concepts of the DMC control algorithm were originally presented by Cutler and Ramaker (1979). The linear DMC algorithm uses time-domain step response models (*convolution* models) to estimate future changes in the manipulated variables that lead the controlled variables to follow an optimal path (set point). This optimal path is determined by optimizing an objective function.

The controller objective is to determine a set of control actions in the future to minimize the objective function, which is composed of two parts:

1) The difference between the setpoint and the model predictions of the process within a prediction horizon (NP).
2) The sum of the squared changes in the manipulated variable multiplied by the square of the suppression factor in a control horizon (NC)

The second part is used to restrict the movements of the manipulated variable.

The objective function is the following (Luyben, 1990):

$$J = \sum_{i=1}^{NP} \left[ysp - \hat{y}_{CL,i} \right]^2 + f^2 \sum_{i=1}^{NC} \left[(\Delta u_k)^{new} \right]^2 \tag{1}$$

where:

ysp= set point, f = weighting factor (suppression factor) which prevent large swings in the manipulated inputs, Δu = increment in the manipulated variable, $(\Delta u)^{new}$ = value of the future changes in the manipulated variable that minimize the performance index *J*.

$\hat{y}_{CL,}$ = prediction made by the model on a prediction horizon called "closed loop response" so $\hat{y}_{CL,i}$, is the value at the *ith* step into the future.

$$\hat{y}_{CL,i} = \hat{y}_{OL,i} + \sum_{k=1}^{NC} a_{ik} (\Delta u_k)^{new} \tag{2}$$

$\hat{y}_{OL,}$ = prediction of the open loop response.

The best values of the future changes in the manipulated variables $(\Delta u)^{new}$, such that the performance index, *J,* is minimized, are obtained by the equation:

$$(\Delta u)^{new} = \left[A^T A + f^2 I\right]^{-1} A^T E \tag{3}$$

where: $\left[A^T A + f^2 I\right]^{-1} A^T$ is a matrix of constant elements. A is the matrix of coefficients of the step response. E is the matrix of elements $e_i = ysp - y_{OL,i}$, where i=1,2…NP.

Minimization of the objective function, Eq. (1), is subject to certain restrictions on manipulated and controlled variables.

$$y_{\min} \le \hat{y}_{k+i} \le y_{\max} \qquad (i = 1,\ldots,NP) \tag{4}$$

$$u_{\min} \le u_{k+i} \le u_{\max} \qquad (i = 1,\ldots,NC) \tag{5}$$

$$\left|u_{k+i} - u_{k+i-1}\right| \le \Delta u_{\max} \qquad (i = 1,\ldots,NC) \tag{6}$$

The DMC uses a corrected value for the prediction ŷ in equation (1), incorporating a feedback strategy. At instant *k*, the predicted value of the output is compared to the measured value; the difference is used to correct the predicted value ŷ in the NP moments ahead. For example, for the moment *k*+1:

$$y^c_{k+1} = \hat{y}_{k+1} + (y_k - \hat{y}_k) \tag{7}$$

Although the optimization is based on a control horizon, only the first control action is implemented and the optimization problem is solved again at the next sampling time. The DMC algorithm has the following steps at each point in time (Luyben, 1990):

1) Calculate the NP values of $\hat{y}_{OL}$, using the past values of the manipulated variables and the present measured value of the controlled variable.
2) Calculate the NC values of the future changes in the manipulated variables, $(\Delta u_k)^{new}$ from Eq. (3).

3) Implement the first change $(\Delta u_1)^{new}$.
4) At the next sampling period, measure the controlled variable to get a new value and repeat the steps above.

2. Extractive Fermentation Process for Bioethanol Production

Figure 1 shows the flow sheet of the extractive fermentation process for ethanol production using sugar cane molasses as substrate and the yeast *Saccharomyces cerevisiae* in the fermentation process for conversion of sugar into alcohol.

The scheme couples the fermentor to a flash distillation system under vacuum. The flash evaporator is equivalent to a stage of a distillation column. The system operates at reduced pressures (vacuum) facilitating the separation of ethanol- water mixture at low temperatures.

The extractive fermentation system has the following equipment: fermentor, filtration system, flash tank, helical pump, peristaltic pump, vacuum system and condensing system. The process output variable to the proposed control system is the fermentor ethanol concentration (P), and the input variables considered to develop the plant model based on ANN are the following:

1) Permeate flow rate (F_P)
2) Fermentor purge flow rate (F_{PU})
3) Feed flow rate to the fermentor (F_0)
4) Fermentor biomass concentration (X)
5) Fermentor substrate concentration (S)
6) Residence time (tr)
7) Fermentor Glycerol concentration (G)
8) Cell viability (C)
9) Fermentor outlet flow rate (F)
10) Fermentor temperature (T_{ferm})
11) Flash tank inlet flow rate (F_{in})
12) Flash tank temperature (T_{flash})
13) Flash tank pressure (P_{flash})

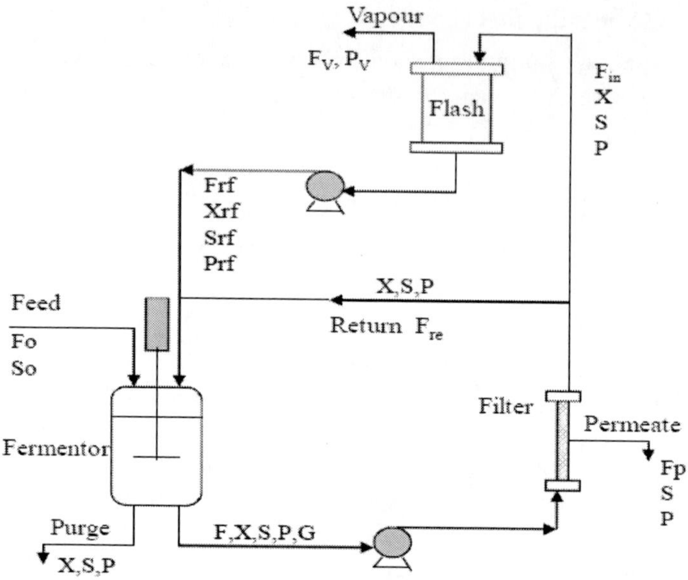

Figure 1. Flow diagram of the ethanol production pilot plant with extractive fermentation process.

3. Plant Model Based on ANN

3.1. ANN Configurations

Modeling of biotechnological processes is complex, mainly due to the difficulties in determining accurate kinetic models. Neural networks can be used as a rapid and accurate modeling technique for these processes, avoiding the choice among many possible kinetic models (Ccopa Rivera et al., 2010 a and b). The alcoholic fermentation, nevertheless, is a well-studied process and its kinetics has been described by many authors (Atala et al., 2001; Andrade et al, 2007). The difficulty in describing the process studied in this chapter with a phenomenological mathematical model is the modeling of the flash tank. Most of the substances found in the culture broth to be separated in the flash tank are polar components. Besides, the very low concentration of many other components (apart from ethanol and water), called congeners, lead to difficulties in correlating and predicting the concentration of the distilled product in the continuous extractive fermentation (Álvarez et al., 2008).

In this study a multilayer feedforward network was proposed to model the extractive alcoholic fermentation process, with a tan–sigmoid transfer function on the hidden layer and a linear transfer function on the output layer. The Levenberg-Marquardt method for backpropagation training was used. The data set of inputs and targets was normalized before training to make it more efficient, in the range [-1,1] as follows.

$$x_{iN} = \frac{2*(x_i - x_{\min})}{x_{\max} - x_{\min}} - 1, \qquad (8)$$

where: $x_{\min}$, $x_{\max}$ and x_{iN} are minimum, maximum and normalized values of the variable x_i.

The input, x_i, weights, W_{ji}, bias, b_j^1, transfer functions, f, and the value of the output variable ge_k^O are related by the following equation:

$$ge_k^O = f^2\left(\sum_{j=1}^{N} W_{kj} f^1\left(\sum_{i=1}^{M} W_{ji} x_i + b_j^1\right) + b_k^2\right) \qquad (9)$$

where: j = 1,…, N; i = 1,…, M ; k = 1,…, K

The weight W_{ji}, have a source on the *i*th input (x_i), and a destination on the *j*th neurons on the hidden layer (layer 1). The weight W_{kj}, have a source on the *j*th neuron of the hidden layer, and a destination on the *k*th neurons of the output layer (layer 2). Superscript 1 indicates that the bias *b*, or the activation function *f* is associated with the first layer. Superscript 2 indicates layer 2. The glucose estimated in the *k*th neuron of the output (*o*) layer is represented by ge_k^O.

The training procedure requires a set of inputs and outputs, so that a particular input leads to a specific target output. During the training process, weights and biases are iteratively adjusted to minimize a performance function (*F*), which is expressed in terms of the mean of squared error (*mse*) that exists between the neural network estimated output and the specified targets.

The information used to train the ANN based model of the fermentation process corresponds to an extractive fermentation experiment that had a total operating time of about 400 hours. The data were stored by the monitoring program of the pilot plant.

3.2. ANN Model Selection

The number of neurons in the hidden layers was determined by using the cross-validation technique in order to avoid model overfitting and to achieve good generalization from the training data set. This technique splits the data sample into a training data set and a validation dataset. The performance of the trained neural network is evaluated by the ability to predict the elements of the validation dataset, which is expressed in terms of mean square error (mse).

$$mse = \frac{1}{N}\sum_{k=1}^{N}\left(t_k - ge_k^O\right)^2 \quad (10)$$

In equation (10), k is the number of data points in the validation data set which vary between 1 and N, t_k is the kth *target value*, ge_k^O is the kth *net estimated output value.*

In the present work 75% of the data were used for training and 25% of the data which was unknown by the network was used for validation. Neural networks of configurations 13: N: 1, where N is the number of neurons in the hidden layer (N Є [4, 11]), were tested in order to select the best neural network based model of the process. It was selected a neural network topology 13: 7: 1, with 7 neurons in the hidden layer because this configuration presented the lowest values of *mse* in the validation test. Figure 2 shows the ANN selected model topology. Table 1 presents the *mse* and R^2 values obtained after the validation test with a different number of neurons in the ANN hidden layer.

Table 1. mean squared error (mse) and R^2 values in the validation test

Neurons	mse	R^2
4	0.76560	0.9935
5	0.04548	0.9962
6	0.03274	0.9973
7	0.01489	0.9988
8	0.05223	0.9978
9	0.06690	0.9945
10	0.02585	0.9979
11	0.05760	0.9953

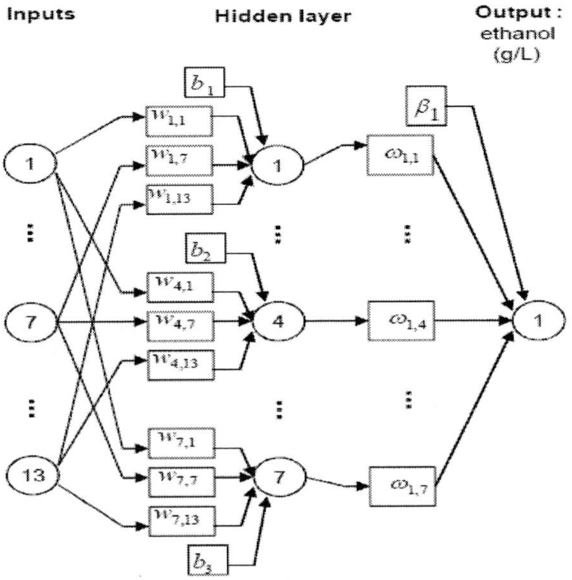

Figure 2. ANN selected model topology.

3.3. Validation Parity Plot

Figure 3 presents the parity plot of the predicted values with the selected ANN and the experimental values of the validation set. The results show that the selected neural network presented a good performance in predicting target values in the validation test.

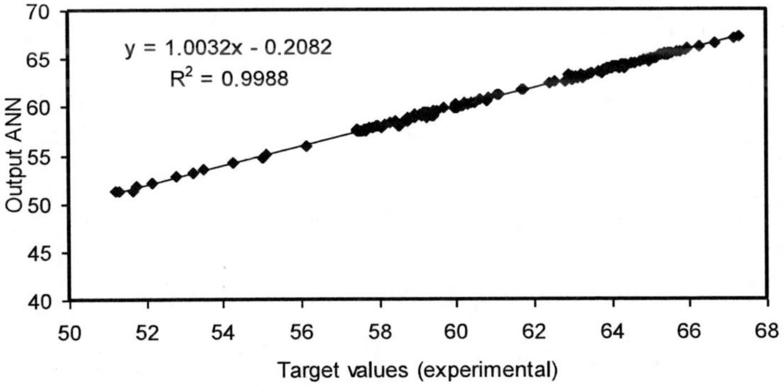

Figure 3. Parity plot of the predicted values with the selected ANN and the experimental values.

Figure 4 shows that the nonlinear experimental behavior of the ethanol concentration in the fermentor of the pilot plant is modeled accurately by the selected ANN.

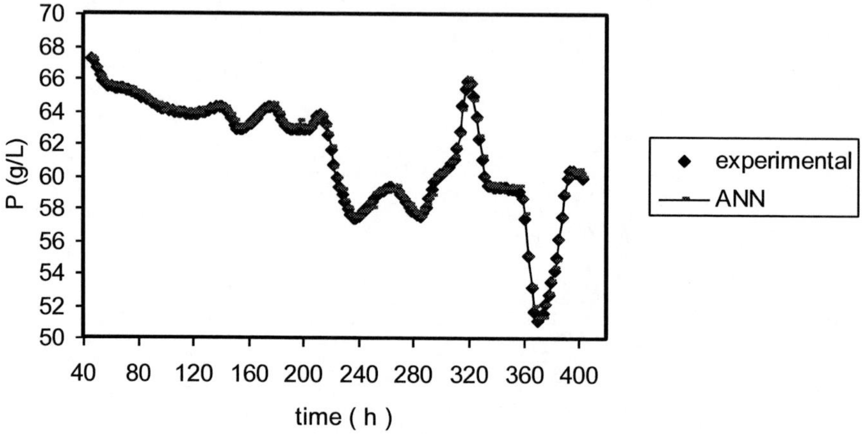

Figure 4. Experimental and ANN predicted values of the fermentor ethanol concentration in the pilot plant with extractive fermentation process.

4. Use of Variable Contributions to the ANN Output to Identify the Importance of Variables

To develop the predictive controller, the first step before training the network to be used as an internal model is to define the inputs and outputs, i.e. what will be the manipulated and controlled variables. In the present chapter the method of transfer functions (Boza and Costa, 2010) was used in order to determine the importance of the effect of each input variable on the output variable. This method uses the interconnection weights of multilayer feedforward neural networks to extract information about the cause and effect relationship between input and output variables. The input-output dependencies are expressed in terms of transfer functions, which are used with block diagrams to estimate the importance of the influence of the inputs on the outputs.The results are shown in Table 2.

Table 2. Input variable contribution to the output variable (ethanol concentration, P), using the transfer function method

Order	Variable	Importance
1	Permeate flow rate (F_P)	0.2071
2	Fermentor purge flow rate (F_{PU})	0.2033
3	Feed flow rate to the fermentor (F_0)	0.1904
4	Fermentor biomass concentration (X)	0.1205
5	Fermentor substrate concentration (S)	0.0772
6	Residence time (tr)	0.0716
7	Fermentor glycerol concentration (G)	0.0571
8	Cell viability (Cv)	0.0484
9	Fermentor outlet flow rate (F)	0.0112
10	Fermentor temperature (T_{ferm})	0.0079
11	Flash tank inlet flow rate (F_{in})	0.0038
12	Flash tank temperature (T_{flash})	0.0011
13	Flash tank pressure (P_{flash})	0.0004

Grouping the variables mentioned in Table 2 we have: a) Variables that can be manipulated in control loops: F_P, F_{PU}, F_0, tr, F; b) Variables that cannot be controlled in control loops because their values are obtained by off-line analysis: X, S, G, C; and c) Variables that can be controlled (all of them have control loops in the pilot plant): T_{ferm}, F_{in}, T_{flash} and P_{flash},. Among the possible manipulated variables F_P, F_{PU}, F_0 and tr have the higher influence on the variable to be controlled, the concentration of ethanol in the fermentor.

5. Use of Random Step Disturbances to Select the Manipulated and Controlled Variables

Profiles of the ethanol concentration in fermentor g/L (output) **were** obtained by performing random step changes in each one of the possible manipulated variables. Then the profiles of both input changes and output responses were compared in order to determine by similarity the manipulated variable which contributes more directly to the behavior of the output variable. Figures 5 to 9 show that best variables to be manipulated to control ethanol concentration in the fermentor are the purge stream flow rate (F_{PU}) and the residence time (tr), because changes in these input variables led to changes directly related in the output variable.

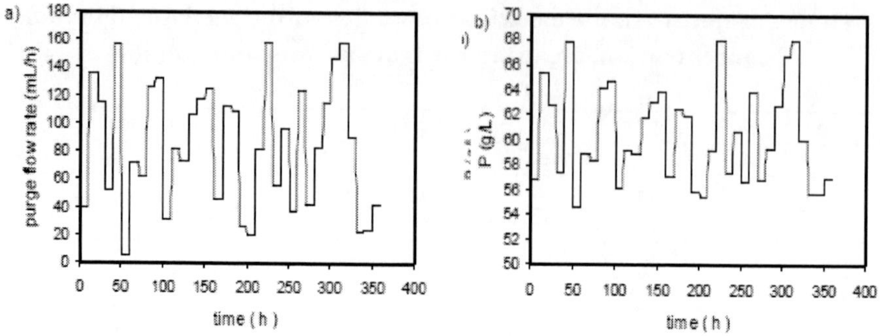

Figure 5. a) Input: random step disturbances in purge flow rate. b) Output: Fermentor ethanol concentration (P).

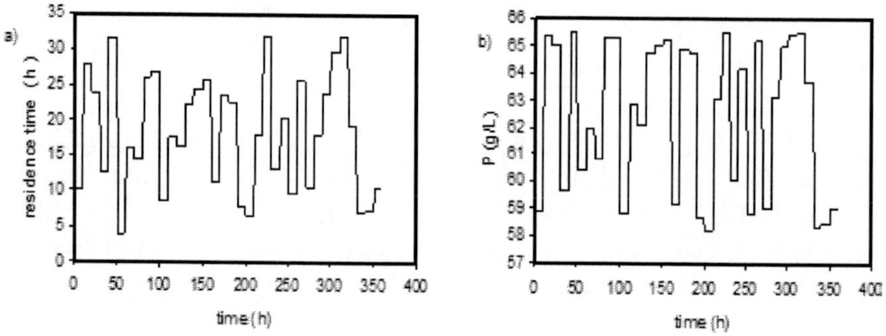

Figure 6. a) Input: Random step disturbances in residence time (tr) b) Output: Fermentor ethanol concentration (P).

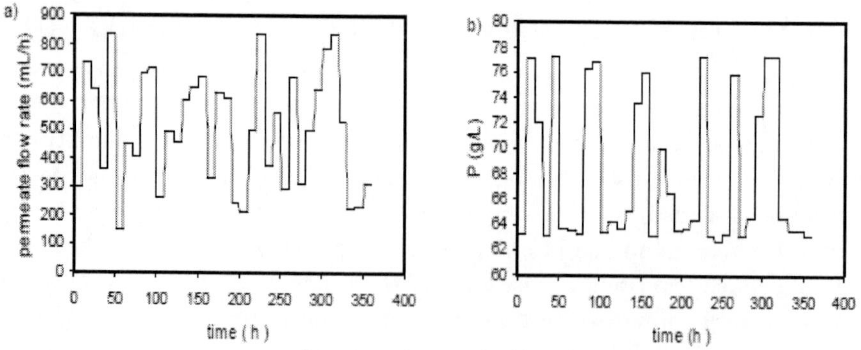

Figure 7. a) Input: Random step disturbances in permeate flow rate. b) Output: Fermentor ethanol concentration (P).

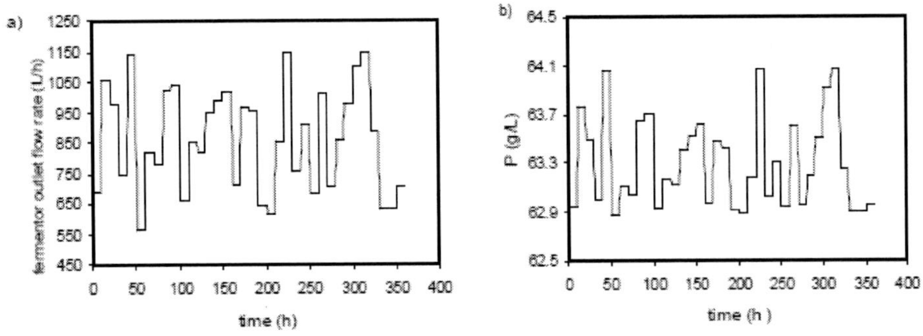

Figure 8. a) Input: Random step disturbances in fermentor outlet flow rate. b) Output: Fermentor ethanol concentration (P).

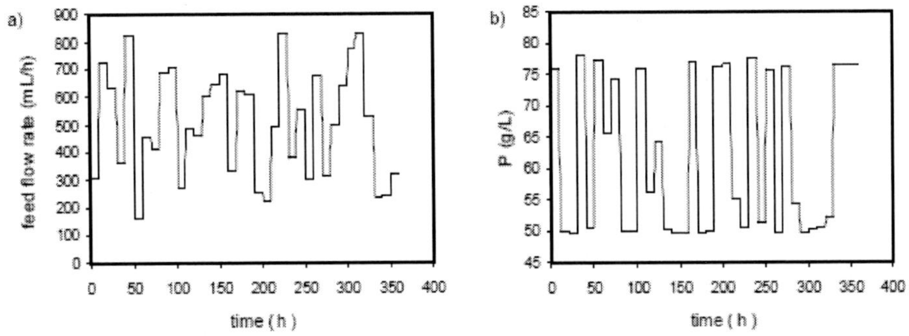

Figure 9. a) Input: Random step disturbances in feed flow rate to the fermentor. b) Output: Fermentor ethanol concentration (P).

Figure 8 shows that the behavior of the fermentor outlet flow rate also contributes directly to the behavior of the output variable P. However, variation on the input led to a very small range of variation of values of ethanol concentration, as can be seen in Figure 8b. Using the analysis of responses to random step disturbances, the selected variable to be manipulated to control the fermentor ethanol concentration P was the purge flow rate (F_{PU}). Thus, the developed nonlinear controller aims to control the concentration of ethanol in the fermentor (P), manipulating the fermentor purge flow rate (F_{PU}).

6. NETWORK TRAINING

In the present work the internal model of the DMC controller is a network with two inputs: the manipulated and controlled variables in the current time and one output: the controlled variable a step forward; as was proposed in several works of nonlinear predictive control in the literature (Costa et al., 2002, 2003; Santos et al., 2000). The same authors suggest that the training data should be generated by making random step disturbances in the manipulated variable.

To generate data to train the network 36 successive random step disturbances were made in the manipulated variable, at each 15 h, as shown in Figure 10. The interval between disturbances was chosen to ensure that new steady states were reached, as suggested by Santos et al. (2000).

The ANN used as internal model has a topology 2:4:1, with 4 neurons in the hidden layer, the performance in describing the training data set was measured by the following equation (Milton and Arnold, 1990):

$$corr = (1 - \frac{SEE}{S\tau\tau})100\% \qquad (11)$$

where $SEE = \sum_{k=1}^{N}(y_e(k) - y(k))^2$, $S\tau\tau = \sum_{k=1}^{N}(y_e(k) - \overline{y_e})^2$, $y_e(k)$ is an experimental output k, y(k) is the corresponding ANN output, $\overline{y_e}$ is the experimental output average and N is the number of experimental data.

The network described the training data with a correlation (*corr*) of 99.21% is calculated by equation (11). Figure 10 shows the results.

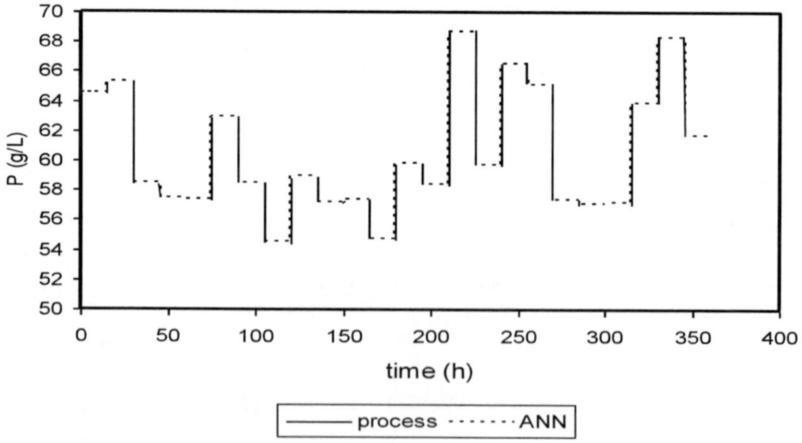

Figure 10. Measured values of fermentor ethanol concentration (g/L) and ANN predicted values after training.

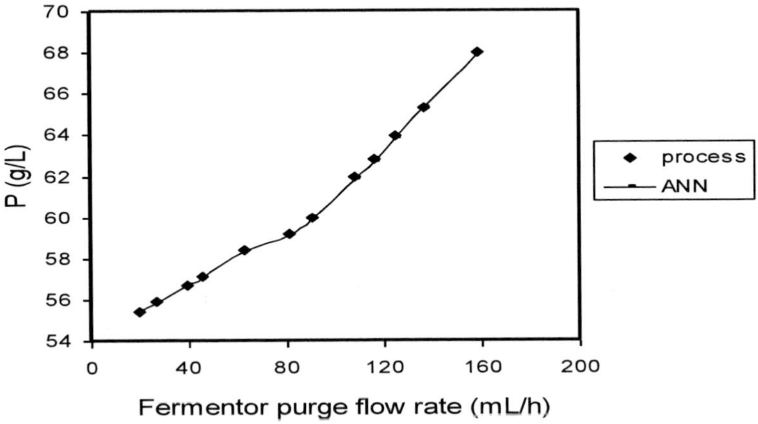

Figure 11. Performance of the neural network in predicting the steady states of the process.

Another important step in the network training is the evaluation of their predictive ability. Usually when the network is used to identify processes, their predictive ability is tested using a set of validation data, i.e. a set of input data that have not been used for training. Fonseca (1998), however, mentions that the networks often yield satisfactory results in the validation test but do not perform well in control applications. This author suggests that the network should be evaluated for its ability to predict the steady states of the plant. Other authors used the same approach to evaluate the performance of networks used as models for predictive control schemes (Santos et al., 2000; Costa et al., 2002, 2003). Thus, this is another test performed to evaluate the predictive ability of the network in this chapter. The network was used to predict the steady states of the fermentation process for various values of purge stream flow rate. The results are shown in Figure 11. We can observe that the network was able to provide a good description of the steady states of the process.

7. Nonlinear Predictive Control

A SISO nonlinear predictive control scheme was developed to control the ethanol concentration in the fermentor manipulating the purge flow rate. The internal model neural network was trained using the current values of manipulated and controlled variables as input and the controlled variable one step ahead as output. As the network is trained to predict only one step ahead, in the control scheme developed, the network is iterated for a total of NP predictions in the future, where NP is the prediction horizon. Thus, the output of the same network is used as input in the subsequent iterations. The sampling time considered was 10 minutes. Constraints on manipulated and controlled variables were the following:

$P_{k+1} \geq 0$ (i = 1,...., NP)

$5.7 \leq F_{PU} \leq 180$ (i = 1,...., NC)

$\left| (F_{PU})_{k+i} - (F_{PU})_{k+i-1} \right| \leq 50$ (i = 1,...., NC)

The controller parameters were chosen after trial and error tests:

$f = 0.18$, NP = 10 and NC =1.

To assess the controller performance rigorous tests were performed by changing the setpoint and the permeate flow rate (F_P). This variable was chosen as a disturbance because it is the variable that most influences the controlled variable besides the manipulated variable F_{PU} (see Table 2).

- Step disturbance in the permeate flow rate.

In this test step disturbances of ± 10% were made in the permeate flow rate (F_P). The results are shown in Figs. 12 and 13.

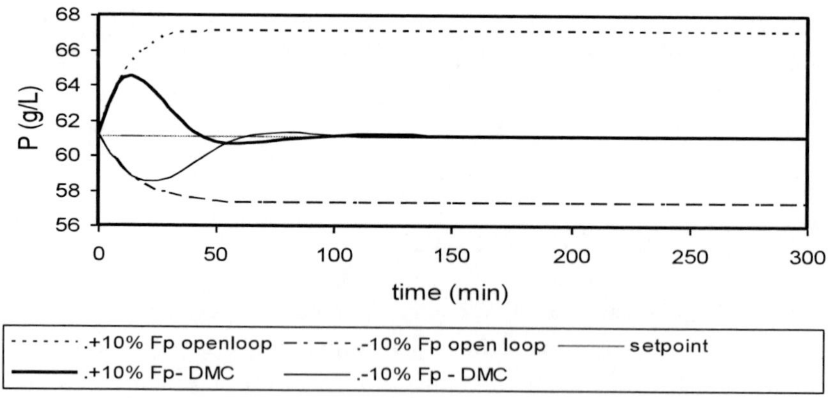

Figure 12. Controlled variable responses against time for step disturbance of ± 10% in the permeate flow rate (Fp).

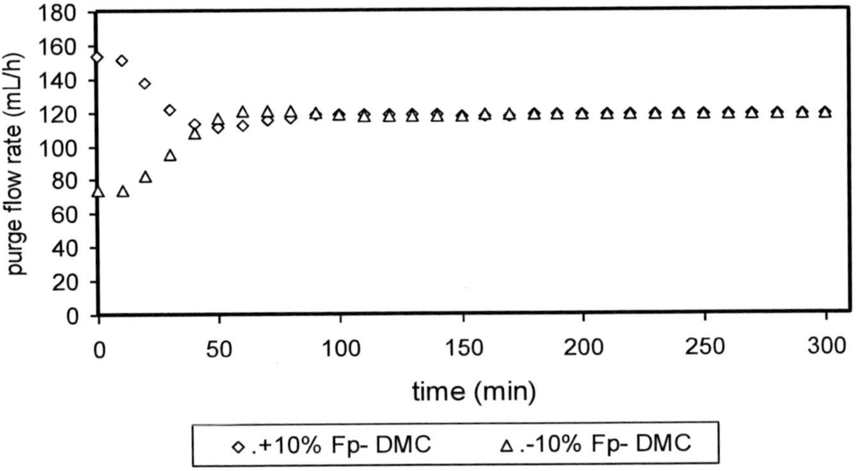

Figure 13. DMC controller actions to reach the steady state for an initial step disturbance of ± 10% in the permeate flow rate (Fp).

Figures 12 and 13 show the behavior of controlled and manipulated variables with time. In Figure 12 the system response in open loop to the same disturbances is also shown. It may be noted that the controller was able to bring the system back to steady state.

- Successive step disturbances in the set point.

In this case successive step disturbances were made in the value of the set point. Figures 14a and 14b show the behavior of controlled and manipulated variables, respectively. It can be observed that the controller performed well bringing the system back to a steady state.

- Combined set-point changes and step disturbances on the permeate flow rate (Fp).

In this case simultaneous changes on set point and step disturbances on Fp were made. Figure 15 shows that the controller showed good performance.

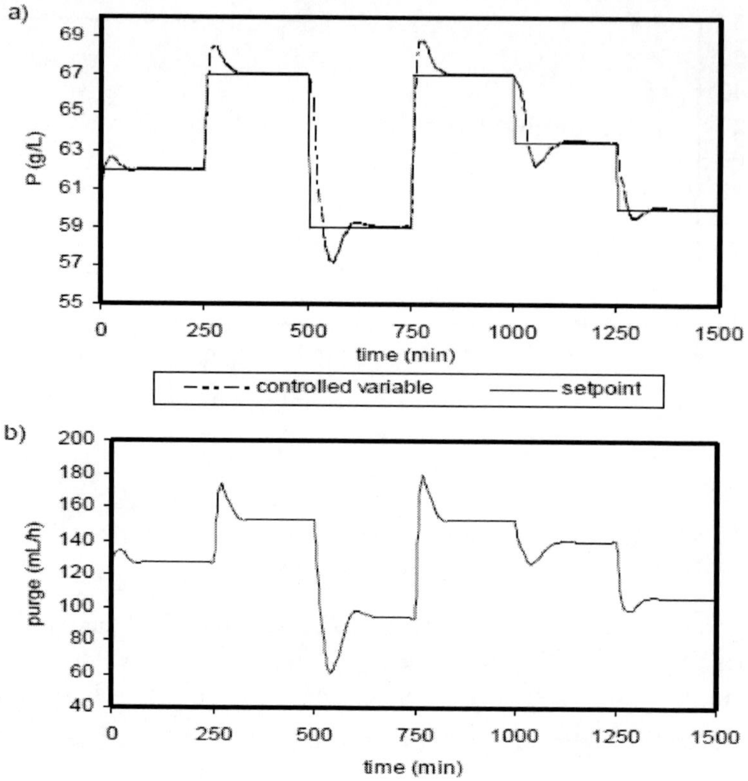

Figure 14. Controller behavior for **set-point** changes: a) Controlled variable responses. b) Manipulated variable changes.

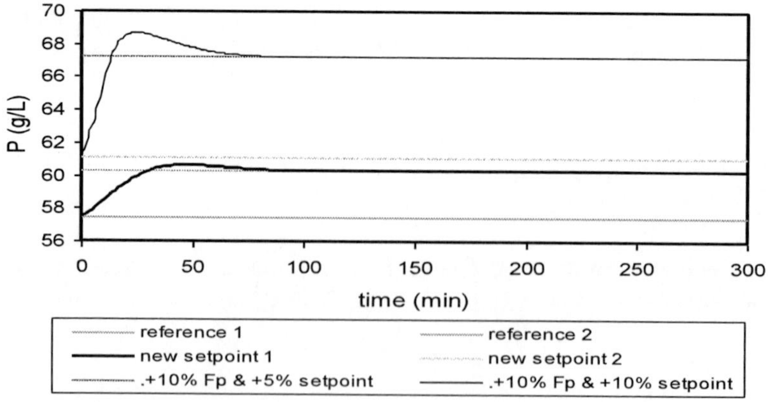

Figure 15. Controller responses for set-point changes and step disturbances on permeate flow rate (Fp).

- Effect of the weighting factor f, on the damping ratio of the closed loop response of the DMC controller with ANN based internal model.

Figure 16 shows that the smaller the weighting factor f, the bigger the changes in the controlled variable, so the controller response is faster but with increased oscillation. The bigger the weighting factor f, the bigger the damping effect on the response, so the controller response is slower against perturbations. In these tests the prediction horizon (NP) and the control horizon (NC) remained constant.

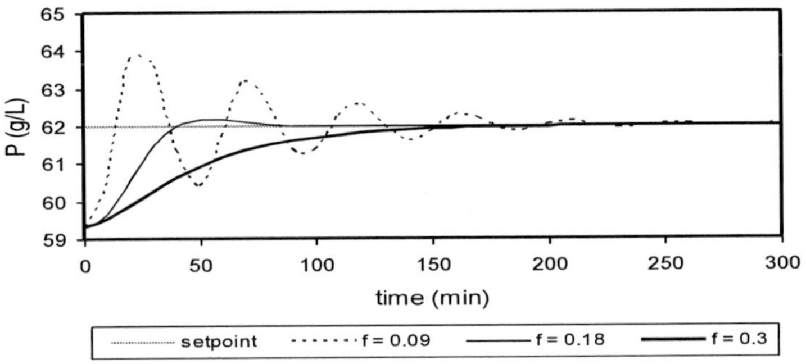

Figure 16. Damping effect for different values of f the weighting factor for Δu's in the controller objective function.

Figure 17 shows that the smaller the weighting factor f, the bigger the changes in the Δu's, and the smaller the damping effect in the response of the controller.

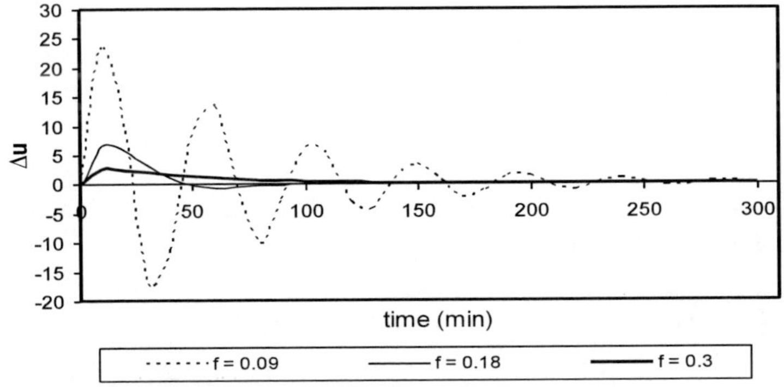

Figure 17. Changes in the manipulated variable for different values of f the weighting factor for Δu's in the controller objective function.

Conclusion

In this chapter a nonlinear control algorithm was developed to control the ethanol concentration in the fermentor of a continuous extractive fermentation process using a feedforward neural network as the internal model in the control strategy based on DMC.

The feedforward network was shown to be suitable for its use as an internal model of the nonlinear controller. One of the most important stages in the development of the nonlinear controller was the development of the internal model. The choice of inputs and outputs of the network, how to generate the set of training data and how to evaluate the predictive ability of the network were very important to the performance of the nonlinear controller. The choice of a network whose inputs are the manipulated and controlled variables in the current time, and the output is the controlled variable one step ahead was adequate in the case study considered. The developed nonlinear controller showed good control performance in all tests that were applied.

References

Andrade, R.R.; Ccopa Rivera, E.; Costa, A. C.; Atala, D. I. P.; Maugeri, F.; Maciel Filho, R. *Appl. Biochem. Biotechnol.* 2007, *136 - 140*, 753 - 763.

Álvarez V.H.; Ccopa Rivera, E.; Costa, A. C.; Maciel Filho, R.;Wolf Maciel M.R.; Aznar, M. *Appl. Biochem. Biotechnol.* 2008, *148*,141-149.

Atala, D. I. P.; Costa, A. C.; Maciel Filho, R.; Maugeri, F. *Appl. Biochem. Biotechnol.* 2001, *91-93*, 353 - 365.

Boza, E.; Costa, A. (2010). The method of Transfer Functions and Block Diagrams to study the contribution of variables in artificial neural networks used for chemical and bio-process modeling. Manuscript submitted for publication.

Ccopa Rivera, E.; Atala, D. I. P.; Maugeri, F.; Costa, A. C.; Maciel Filho, R. *Chem. Eng. Process.* 2010, *49*, 402-409.

Ccopa Rivera, E.; Rabelo, S.C.; Reis Garcia, D.; Maciel Filho, R.; Costa, A. C. *J. Chem. Technol. Biotechnol.* 2010, *85,* 983 - 992.

Chu, J.; Su, H.; Gao, F.; Wu, J. *Annu. Rev. Contr.* 1998, *22*, 59-72.

Costa, A. C.; Atala, D. I. P.; Maugeri, F.; Maciel Filho, R. *Process Biochem.* 2001, *37(2)*, 125-137.

Costa, A.C.; Meleiro, L.A.C.; Maciel Filho, R. *Process Biochem.* 2002, *38(5)*, 743-750.

Costa, A.C.; Maciel Filho, R. *Can. J. Chem Eng.* 2003, *81*, 1109-1118.

Dechechi, E.C. *Predictive adaptative advanced control "DMC Multivariavel Adaptative"*; (PhD thesis) UNICAMP: Campinas - SP, 1998.

Fonseca, E.F. (1998). *Predictive control based in neural networks of continuous alcoholic fermentation*; (M.S. Thesis) UFPE, Recife - PE. 1998.

Gattu, G.; E. Zafiriou. *Chem. Eng. J.* 1999, *75*, 21-29.

Gawkowski, P.; Lawrynczuk, M.; Marusak, P.; Sosnowski, J.; Tatjewski, P. In *Proceedings of the International Multiconference on Computer Science and Information Technology*; Ganzha, M; Paprzycki, M; Pelech-Pilichowski; Wisla, Poland, 2007; *Vol 2*, pp 903–912.

Henrique, H.M. *A Contribution to the Study of Neural Networks Applied to Process Control*; (PhD thesis) PEQ/COPPE/UFRJ, Rio de Janeiro, RJ, 1999.

Hussain, M.A. *Artif Intell En.* 1999, *13*, 55-68.

Kambhampati, C.; Mason, J.D.; Warwick, K. *Automatica.* 2000, *36*. 485-495.

Le Lann, M.V.; Cabassud, M.; Casamatta, G. *Annual Reviews in Control.* 1999, *23*, 25-34.

Lima, N. M. N. Modeling and predictive hybrid control by fuzzy logic for polymerization processes; (M.S. Thesis) UNICAMP: Campinas - SP, 2006.

Luyben, W. *Process modeling, simulation and control for chemical engineers*; Second Edition; McGraw-Hill: New York, NY, 1990, pp 287.

Meleiro, L. A. C.; Filho, R. M.; Campello, R. J. G. B.; Amaral, W. C. In Application of Neural Network and Other Learning Technologies in Process Engineering; Mujtaba, I. M. and Hussain, M.A., Imperial College Press. UK, 2001.

Milton, J.S.; Arnold, J.C. *Introduction to Probability and Statistics*; McGraw Hill, New York, NY, 1990.

Morari, M.; Lee, J.H. *Comput. Chem. Eng.* 1999, *23*, 667- 682.

Nagy, Z, K. *Chem. Eng. J.* 2007. *127(1-3)*, 95-109.

Piché, S.; Sayyar-Rodsari, B.; Johnson, D.; Gerules, M. *IEEE Contr. Syst. Mag.* 2000, *20(3),* 53-62.

Qin, S.J.; Badgwell, T.A. In *Proceedings of the 5th International Conference on Chemical Process Control*; Kantor, J.C. ; Garcia, C.E.; Carnahan, B.; *AIChE Symposium Series No.316*; Tahoe City, CA, 1997; *Vol. 93*, pp 232-256.

Rawlings, J.B. Tutorial overview of model predictive control. *IEEE Contr. Syst. Mag*. 2000, *20(3)*, 38-52.

Rohani, S.; Haeri, M.; Wood, H.C. *Comp. Chem. Eng*. 1999, *23*, 279-286.

Santos, L.O.; Biegler, L.T. *J. Proc. Cont*. 1999, *9*, 233-246.

Tatjewski, P.; Lawrynczuk, M. *Int. J. Appl. Math. Comput. Sci.* 2006,*16 (1)*, 7–26.

In: Industrial Control Systems
Editors: R. Gilbert and A. Schultz
ISBN: 978-1-61209-988-0

Chapter 3

NONDESTRUCTIVE DYNAMIC MONITORING OF ACCELERATED ION BEAMS

A. A. Baldin ***and E .G. Baldina***
LLC "Institute for Advanced Studies "OMEGA",
Dubna, Moscow reg., Russia

ABSTRACT

A complex for dynamic nondestructive diagnostics of accelerated ion beams has been developed. The complex includes the detector, measurement, data collection systems and the software for beam control and optimization of irradiation regimes.

The system of dynamic diagnostics of film irradiation based on transparent (95% transparency) profilometers for track membrane production was developed and implemented at the accelerator complex "Alpha" (JSC "Trackpore Technology", Moscow, Russia). This system provides measurement, dynamic control, and certification of irradiation parameters with a time resolution of 100 μs. Industrial track membrane production regimes with given irradiation density and homogeneity were studied. Practical recommendations on optimal regimes were formulated as part of the metrological control of industrial film irradiation technology.

The system of dynamic monitoring of space-time characteristics of circulating beam based on microchannel plates was developed for the

* e-mail: e.baldina@mail.ru

superconducting accelerator Nuclotron (Joint Institute for Nuclear Research, Dubna, Russia). The system possesses unique time characteristics, it provides measurement of nuclear beam intensities during acceleration in a range from 10^4 to 10^{11} pps.

INTRODUCTION

This paper presents the results of development of the complex of dynamic nondestructive diagnostics of accelerated ion beams for industrial and scientific application.

The task of measuring space-time structure of ion beams both at the output of an accelerator and during acceleration is common in all fields in which charged particle accelerators are used: physics research, industrial, and medical accelerators.

One of the areas of application of industrial cyclotrons is track membrane production. A track membrane represents a polymer film with pores of defined radius formed by specially processed ion tracks; they are used in ecology, biology, medicine, in particular as the material for blood filters in plasmapheresis. The most important parameters of track membranes are membrane strength, irradiation density, and pore size. For medical track membranes nominal values of these parameters usually should not vary by more than 5-10% all over the membrane. The specific features of physical and chemical processing of irradiated film to produce track membranes result in complex coupling of these parameters which requires monitoring of all stages of membrane production technology. The first stage of this technology is the film irradiation with an accelerated ion beam. The most important factors to be controlled at this stage are the irradiation density (number of ions per unit area) and homogeneity (deviations of irradiation density from the nominal value toward higher values may result in lower membrane strength, and toward lower values, in lower throughput capacity, both being unacceptable in medical blood filters). Section 1 of this study is devoted to the system of dynamic monitoring of accelerated ion beam at the output of the accelerator for film irradiation. Section 2 is devoted to the monitoring system for a circulating ion beam based on microchannel plates.

1. Dynamic Nondestructuve Beam Diagnostics for Industrial Cyclotron

The dynamic beam monitoring system [1] is designated for online control of film irradiation regime; space-time beam characteristics measured by this system provide online monitoring of irradiation quality which makes it possible to evaluate the properties of track membranes produced from this film. The developed system was implemented at JSC "Trackpore Technology" (Dubna, Russia), full-cycle manufacturer of blood filters from track membranes.

1.1. Simulation

Let us study the influence of different irradiation regimes on film irradiation homogeneity.

The problem is formulated as follows. The beam is characterized by the source function $f(x, y, t)$ determining the time evolution of particle distribution in the beam cross section (x and y are the horizontal and vertical coordinates, respectively, and t is the time). In order to obtain the ion distribution on the film this source function should be modified by the action of scanning magnets that shift the beam in the horizontal and vertical direction in a known way, and the movement of the film in the vertical direction. This operator is written as $\varphi = \varphi_x \varphi_y$, where $\varphi_x = \varphi_x(\omega_x, h_x)$, (ω_x is the horizontal scanning frequency and h_x is the shifting amplitude in the horizontal direction), $\varphi_y = \varphi_y(\omega_y, h_y, V)$ (ω_y is the vertical scanning frequency, h_y is the vertical shifting amplitude, and V is the film movement velocity in the vertical direction).

Then the ion distribution on the film is determined as $F(x, y) = \varphi f(x, y, t)$. Here, $F(x, y)$ is the function of ion (pore) distribution on the film.

Let us consider a realistic situation of scanning in horizontal direction by a Gaussian beam,

$$f(x,y,t) = \psi(x,y,x_c,y_c)\cdot I(t),$$

where the ion distribution in the beam cross section is described as

$$\psi(x,y,x_c,y_c) = \frac{1}{2\pi r_x r_y}\exp\left(-\frac{(x_c-x)^2}{r_x^{\,2}} - \frac{(y_c-y)^2}{r_y^{\,2}}\right)$$

here, x_c and y_c are the coordinates of the beam central point; the action of the scanning magnet and film movement is described as

$$x_c(t) = x_0 + h_x\sin(2\pi\omega_x t),\; y_c(t) = y_0 + Vt$$

For investigation of the influence of time structure of the beam two cases are considered: a periodic function of time,

$$I(t) = I_0(1 + a\sin(2\pi\Omega t))$$

where I_0 is the nominal beam current amplitude, a is the amplitude variation coefficient, and Ω is the amplitude modulation frequency (for $a = 0$ or $\Omega = 0$ the beam current is constant), and an experimentally measured distribution (see Figure 1).

Let us solve the problem by numerical integration,

$$F(x,y) = \frac{1}{2\pi r_x r_y}\iiint I(t)\psi(x,y,x_c,y_c)dx_c dy_c dt\,.$$

For explicit integration the time step should satisfy the following condition: $\tau < \frac{\min(\Delta x, \Delta y)}{\max(V, 2h_x\omega_x)}$, where Δx, Δy are the integration steps in the horizontal and vertical directions, respectively.

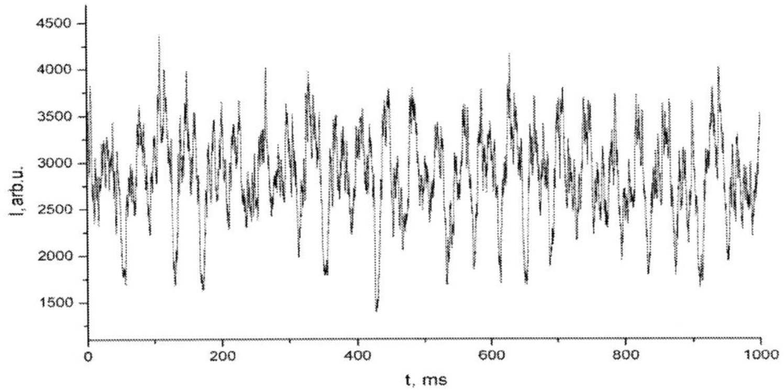

Figure 1. Experimental total beam current measured with a time resolution of 0.2 ms.

The following realistic parameter values were used in the simulation: $r_x = 0.6$ cm, $r_y = 1.2$ cm, $h_x = 20$ cm, $\omega_x = 80$ Hz, $V = 20$ cm/s; and $a = 0.2$, $\Omega = 2.5 \div 50$ Hz for periodic beam current variation.

Simulations showed that the experimentally observed effect of transverse "stripes" of pore distribution on industrially produced track membranes becomes noticeable at ~10 Hz modulation frequency (a period of ~2 cm and an inhomogeneity of ~1.5%); for a modulation frequency of ~5 Hz, the striped character of film irradiation becomes quite pronounced (a period of ~2 cm and an inhomogeneity of ~15%), and is manifested even stronger with decreasing modulation frequency (for ~2.5 Hz we obtain a period of ~4 cm and an inhomogeneity of ~35%).

If the beam size in the vertical direction decreases, the effect of transverse stripes is enhanced; thus, for, for example, for $r_y = 0.598$ cm and $\Omega = 10$ Hz the inhomogeneity reaches 7%, for $\Omega = 5$ Hz, 30%, and for $\Omega = 2.5$ Hz, 44%. With further reduction of the vertical beam size to $r_y = 0.3$ cm transverse inhomogeneity increases to 3, 13, and 30% for 30, 20, and 10 Hz, respectively, with a period of 0.83, 1.25, and 2 cm, respectively. For a rather narrow beam, $r_y = 0.2$ cm, modulation frequencies resulting in noticeable irradiation inhomogeneity are close to 50 Hz often present as the parasitic frequency in power supply networks. In this case irradiation inhomogeneity can reach from 2% in the middle of the film to 10% at a distance of 5-10 cm from the film center toward its edge for moderate modulation coefficient, $a = 0.2$, and realistic film movement velocity, $V = 25$ cm/s, and from 12 to 48%, respectively, for $a = 0.5$, $V = 30$ cm/s.

Film irradiation inhomogeneity can also have the character of "patches" with dimensions depending on the ratio $\frac{V/r_y}{\Omega}$. A simple and efficient method for reducing irradiation inhomogeneity is to decrease the film movement velocity. Thus, even for the worst case of the narrowest beam the inhomogeneity can be reduced to <1% on the whole film width by reducing V by a factor of 2 from 25 to 12.5 cm/s.

Figure 2 illustrates the strong influence of the beam size on the character of pore density distribution on the film.

Let us consider in more detail the dependence of pore density inhomogeneity on different irradiation parameters. We determine the inhomogeneity as

$$\Lambda = \frac{F_{\max} - F_{\min}}{F_{\min}} \cdot 100\,\%,$$

where $F_{\max}$ and $F_{\min}$ are the maximum and minimum densities in the considered region.

a)

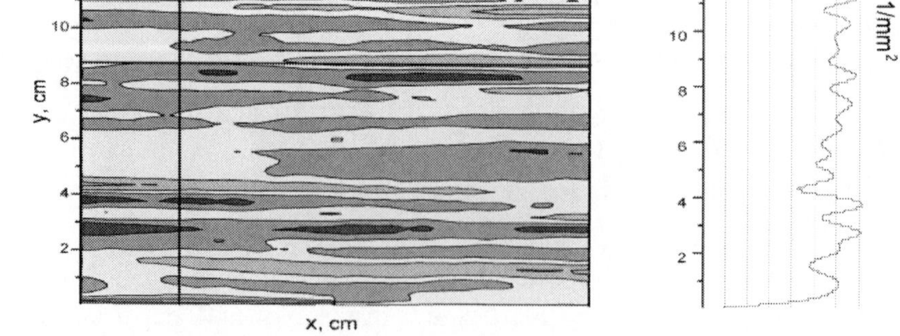

b)

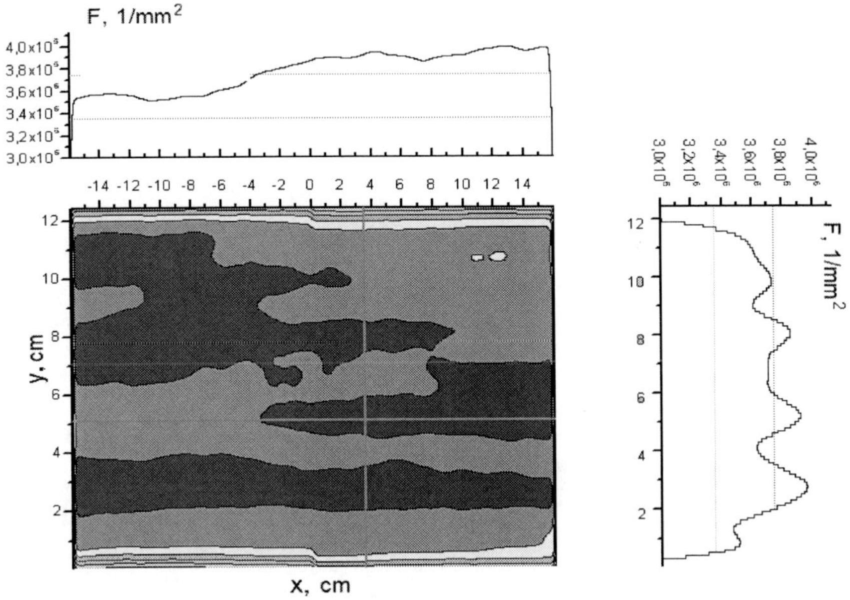

Figure 2. Simulated pore distribution for experimental beam current (Figure 1), $V = 25$ cm/s, $\Omega = 50$ Hz, and (a) $r_y = 0.199$ cm; (b) $r_y = 0.598$ cm. Grey levels show regions with 10% pore density variation.

Figure 3 shows irradiation (or pore density) inhomogeneity as a function of beam current modulation frequency for three modulation amplitudes. It can be seen that the inhomogeneity is the higher, the higher the modulation amplitude, and the lower the modulation frequency. It should be noted that somewhat better result (lower inhomogeneity) for $\Omega = 40$ Hz is explained by the fact that this makes one half of the scanning frequency $\omega_x = 80$ Hz.

Figure 4 shows pore density inhomogeneity as a function of vertical beam size. If the beam is compressed in the vertical direction, a noticeable growth of irradiation inhomogeneity is observed. For a relatively broad beam, $r_y \geq 0.4$ cm (which corresponds to an effective vertical spot size of ≥ 1.2 cm), the inhomogeneity decreases below 10% for all modulation amplitudes under study.

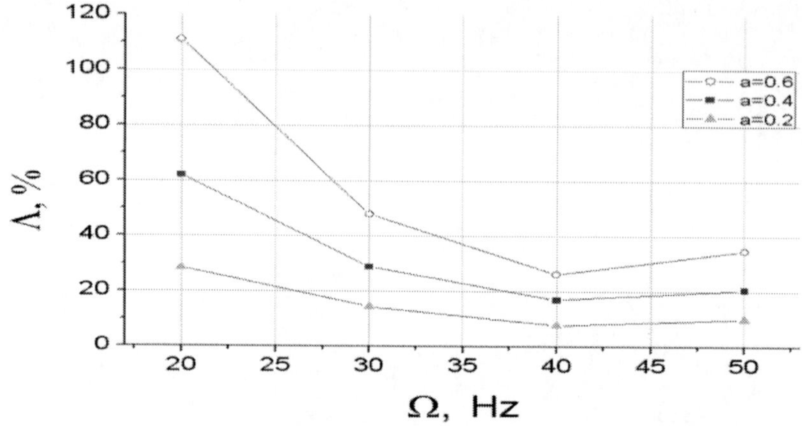

Figure 3. Pore density inhomogeneity Λ at a distance of 10 cm from the film center as a function of beam current modulation frequency Ω for Gaussian beam with $r_x = 0.3$ cm, $r_y = 0.2$ cm, $V = 25$ cm/s and three modulation amplitudes: $a = 0.2,\ 0.4,\ 0.6$.

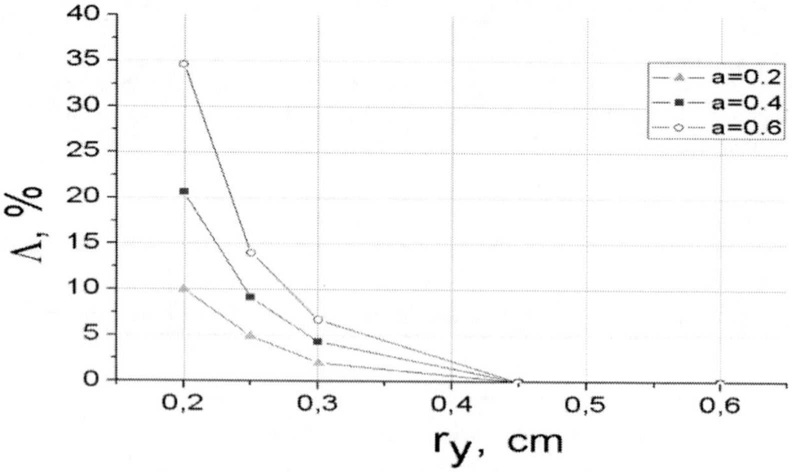

Figure 4. Pore density inhomogeneity Λ at a distance of 10 cm from the film center as a function of vertical beam size r_y for Gaussian beam with $r_x = 0.3$ cm, $V = 25$ cm/s, $\Omega = 50$ Hz and three modulation amplitudes: $a = 0.2,\ 0.4,\ 0.6$.

Figure 5 illustrates the influence of the film movement velocity on the pore density inhomogeneity. It can be seen that this parameter strongly influences the quality of irradiated film, especially in the case of noticeable modulation amplitudes. The film movement velocity can be easily controlled and varied. Therefore, this is the key parameter for tuning irradiation regimes to obtain the desired inhomogeneity. Thus, for example, the reduction of the transport velocity from 25 to 17.5 cm/s (i.e., by 30%) results in Λ reduction by a factor of 3.3–5 (the most pronounced smoothing of the pore density distribution is observed for maximal modulation amplitudes).

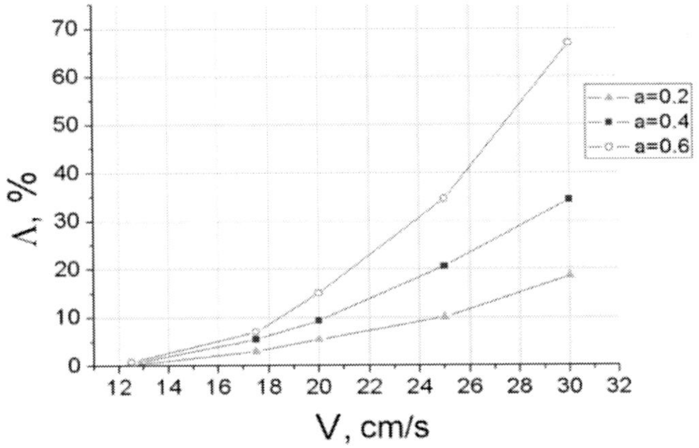

Figure 5. Pore density inhomogeneity Λ at a distance of 10 cm from the film center as a function of film movement velocity V for Gaussian beam with $r_x = 0.3$ cm, $r_y = 0.2$ cm, $\Omega = 50$ Hz and three modulation amplitudes: $a = 0.2,\ 0.4,\ 0.6$.

It should be taken into account that realistic beam current pulsations may considerably exceed $\pm 10\,\%$, reaching a factor of 2 and even more, which naturally increases irradiation inhomogeneity; the instability of the spatial beam position has a similar effect. This demonstrates that the above estimates yield optimistic Λ values.

The numerical study demonstrated that, in order to achieve the required pore density homogeneity, complex adjustment and monitoring of irradiation parameters (smooth ion distribution in the spot with convex cross section and sufficiently large size, $\sim 0.4 \div 2$ cm, time-stable beam current, appropriate film movement velocity, and satisfactory scanning parameters) is required. The most critical parameter is the beam current dynamics (and, partly, the spatial oscillation of the beam spot whose impact can be quite similar). Beam intensity oscillations may result in considerable irradiation inhomogeneities. In order to provide irradiation homogeneity on a level of 5-10% required for medical membrane blood filters, these oscillations should be monitored with a resolution not worse than 0.1-1 ms. The most efficient and feasible method for suppressing these inhomogeneities is to reduce the film movement velocity. In conditions of industrial film irradiation it is desirable to use highest possible film movement velocity in order to efficiently use accelerator beam time; the software solving the problem formulated above in combination with the developed dynamic diagnostic system make it possible to find reasonable balance between production efficiency and membrane quality for particular manufacturer's conditions.

1.2. Dynamic Beam Diagnostic System

The system for dynamic diagnostics of accelerated ion beam consists of the following components: transparent profilometers for measuring the beam current; charge-frequency transducers; PC-compatible readout electronics; and a PC for storing, processing and displaying information. The parameters of the diagnostic system are determined by experimental conditions and required precision of monitoring irradiation characteristics.

For the conditions of industrial film irradiation for track membrane production measured currents are from 100 pA to 1 μA; according to the study above the required time resolution is on the level of 100 μs.

It should be noted that current–voltage transducer amplifiers with typical conversion times on the order of 200 ms, which are often used in such diagnostic systems, are incapable of providing the required time resolution. This inspired the creation of new diagnostic tools.

In the conditions of industrial track membrane production the regime parameters should be monitored permanently during irradiation, rather than at the stage of accelerator adjustment, when destructive measurement methods are acceptable. For this reason measurement should be nondestructive. In order to provide nondestructive beam monitoring high transparency and registration efficiency should be provided. PC-compatible electronics makes the diagnostic system highly adaptable; developing software for data recording, readout, and processing is simpler and less expensive. Also, such a system can be easily extended for including other diagnostic tools and connected to other control systems.

1.2.1. Transparent Profilometers

The technical characteristics of the developed transparent two-coordinate profilometers are as follows:

(i.) isolated stainless steel frame (working region 510 × 90 mm) with a set of contact elements for signal readout;

(ii.) set of grid lamels (wire material 20 μm gold-plated tungsten). Wires were wound with a tension of 21.5 g per wire; the wire orientation was vertical with a step of 2 mm and an inclination angle of 1.3°. Grid lamels were divided into groups with 19 wires in each group and formed 9 independent signal readout groups covering a working range of 360 mm;

(iii.) set of grid lamels at the back side of the frame (wire material 100 μm gold-plated tungsten). Wires were wound with a tension of 30.4 g per wire; the wire orientation was horizontal with a step of 9 mm, covering a working range of 80 mm in the vertical direction and the whole measurement range in the horizontal direction determined by standard width of the irradiated roll films (usually 320 and 600 mm).

The profilometers were installed at the output of the accelerator directly before the irradiated film. The transparency of these profilometers is not worse than 95%, no film shadowing occurs in the course of irradiation.

1.2.2. Charge-Frequency Converters

The new charge–frequency converters used in this measurement system were manufactured based on [2] but with improved noise characteristics and increased sensitivity. The converter contains an integrator with low-input-current (1 pA) AD820 operational amplifier at the input. During the measurement interval *t*, the charge $Q = \int_0^t I(t')dt'$ is received at the input of the integrator from a detector (wire or group of wires). Charge is counted by doses q using the converter commutation circuits, $q = i \cdot \tau$, where i is the current of the stable current source and τ is the digital univibrator pulse duration. Thus, the total charge $Q = N \cdot q$, where N is the number of pulses at the output of the converter; this charge value corresponds to the charge accumulated during the measurement time. This type of conversion yields considerably better noise stability and wide conversion range in the whole interval of intensities of the accelerator complex from an ion source intensity to a minimal intensity of the signal received from transparent profilometers directly before the irradiated sample. These converters were tested in the conditions of total beam blocking by absorbing lamels for matching with destructive total current measurements. Four transducers with the following weighting factors were used: q = 100, 25, 20.6, 20.8 pC/pulse. NIM pulses with a duration of 500 ns at the output of charge-frequency converters allow one to use standard readout electronics (pulse recalculation in a given time interval) both for depicting time beam structure and dynamic feedback with the purpose of stabilizing a given irradiation intensity.

1.2.3. Measurements

Figure 6 shows the average beam position for different scanning regimes. Measurements showed that the beam intensity dynamics was adequately measured by the vertical and horizontal groups of wires of the profilometers. It was established that parasitic pulsations with characteristic frequencies of order of 50 Hz resulting from power source instability are present for all studied scanning regimes from 28 to 130 Hz.

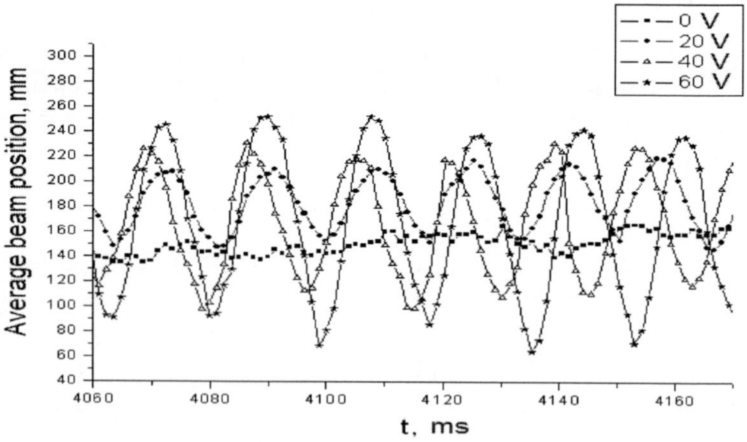

Figure 6. Average beam position as a function of time for different voltages at the scanning magnet.

Figure 7 shows the intensity at the output of the accelerator as a function of average beam intensity measured in the conditions of complete beam absorption with a characteristic averaging time of 200 ms. The intensity variation was achieved by varying the voltage on the buncher. It can be seen that the transparent profilometer preserves linearity in the whole range of measured currents.

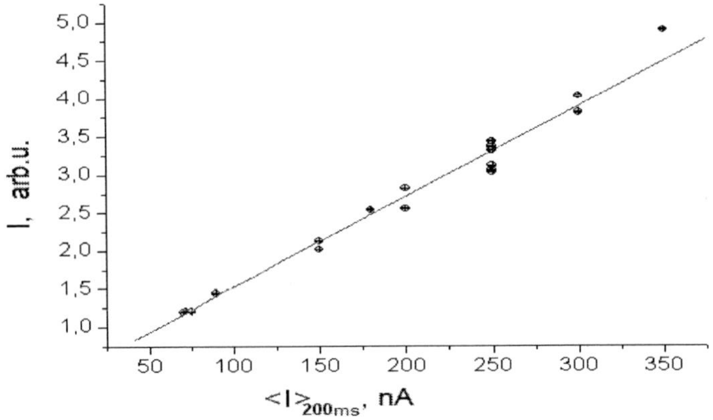

Figure 7. Beam intensity as a function of total beam current averaged over 200 ms.

Operation of scanning magnets and their influence on the readout electronics was studied using specially developed calibration magnetometer based on three-component Hall sensor with a current loop (see Figure 8). The measurements showed that the magnet provides satisfactory scanning linearity at frequencies of ~80 Hz. However, the additional stabilization of the power source and the dynamic monitoring of the magnet operation during irradiation are required.

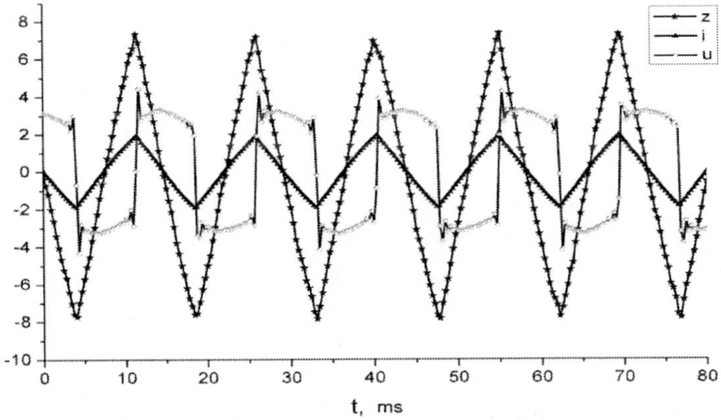

Figure 8. Characteristics of the scanning magnet as functions of time: (*i*) current at the magnet, (*z*) magnetic field, and (*u*) voltage at the current loop.

Figure 9 shows the beam current dynamics in comparison with the signal measured from the current loop placed inside the scanning magnet. Measurements at different scanning frequencies (from 28 to 130 Hz) demonstrated the stability of this diagnostic system to noise from the scanning magnet.

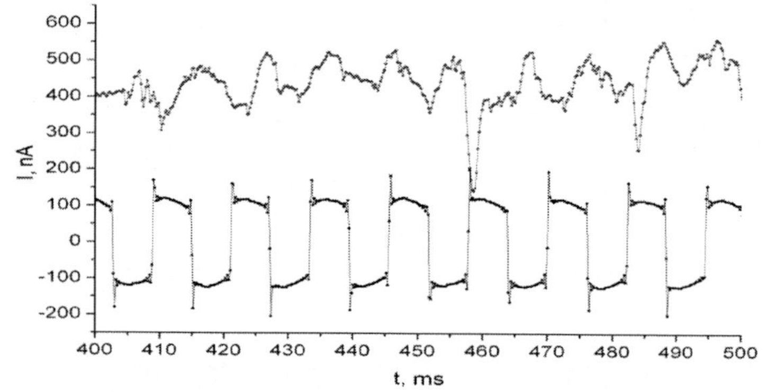

Figure 9. (Upper curve) Beam current and (lower curve) signal from the current loop inside the scanning magnet.

The measurements were performed using the 16-channel ADC module (ZETLab).

For absolute calibration of the developed dynamic beam diagnostic system, the irradiated film was processed to obtain SEM photos of pores which were counted using a specially developed software.

2. Dynamic Nondestructuve Beam Diagnostics for Circulating Beam of Research Accelerator

The basic ion beam parameters to be controlled in acceleration of circulating beams are the beam intensity and spatial profile. It is especially important to monitor the dynamics of these characteristics at beam injection from an ion source to the accelerating ring, acceleration, and extraction to external beam lines. The diagnostic system used for this purpose should satisfy the following requirements: it should provide beam monitoring in a wide range of intensities, possess high spatial resolution in one or, better, two directions, and high time resolution; the system should be nondestructive and operate inside an accelerator ring [3].

Detectors based on microchannel plates (MCP) are widely used in nuclear physics studies. These detectors provide fast time signals and possess sufficient coordinate sensitivity. These detectors were chosen to be used in the dynamic nondestructive diagnostic system for circulating beam of Nuclotron (heavy ion beam accelerator at the Laboratory of High Energy Physics of the Joint Institute for Nuclear Research, Dubna, Russia) [4, 5].

The diagnostic system based on MCP provides reliable registration of circulating ion beam with intensities from 10^4 (at injection energy) to 10^{12} (at an energy of 20 MeV/u and higher) for a vacuum level of $10^{-5}\div10^{-6}$ Pa in a time window from 300 ns to 10 ms with a total registration time during the whole acceleration and extraction cycle up to 15 s.

2.1. Detector Design

The MCP detector consists of two parts (see Figure 10): electrostatic and registering parts. The electrostatic part is designated for transportation of ionized atoms of residual gas under the action of homogeneous electric field. The registering part includes the chevron assembly of two MCPs and the position-sensitive anode. MCPs applied in this detector possess the following characteristics: channel diameter of 0.015 mm, channel inclination angle of 7 @[degree].

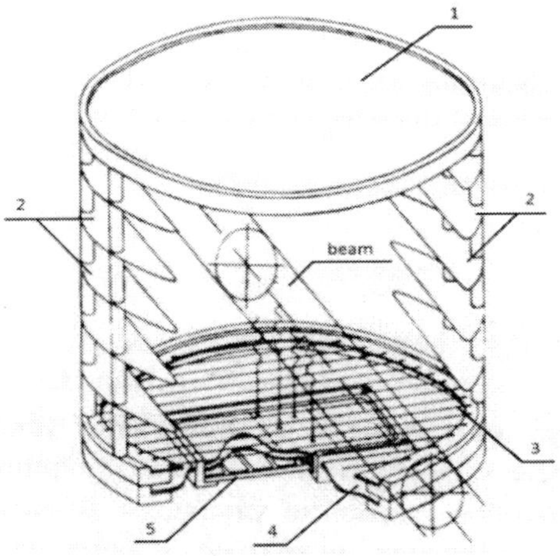

Figure 10. Detector design: 1 - positively charged electrode, 2 - pair electrodes for electrostatic field compensation, 3 - cutting-off grid, 4 - negatively charged split electrode, 5 - MCP pad.

To provide detector transparency, it is installed in the accelerator chamber beyond the circulating beam trajectory. Installation of two perpendicular MCP assemblies makes it possible to measure the beam dynamics, spatially resolved in the plane perpendicular to the beam axis. The spatial resolution is 2mm in each direction. The time resolution is 100ns. The registration region covered by the MCP detector depends on MCP size. In the considered diagnostic system it made 58 and 88 mm.

2.2. Readout Electronics and Measurements

Specially developed multichannel readout electronics is used for signal registration. The data acquisition system provides measurement of signals from MCP pads at given time intervals, thus yielding detailed time-resolved signal dynamics at each pad with a resolution of up to 10 ns. Taking into account typical ion drift times of order of 1 μs, this capability is especially useful for specific acceleration adjustment regimes. Charge collection system of MCP detector can be designed for particular experimental conditions depending on required spatial resolution with account of load characteristics. It should be noted that in this case the detector design does not require essential changes.

Figure 11 shows an example of space-time beam structure at injection.

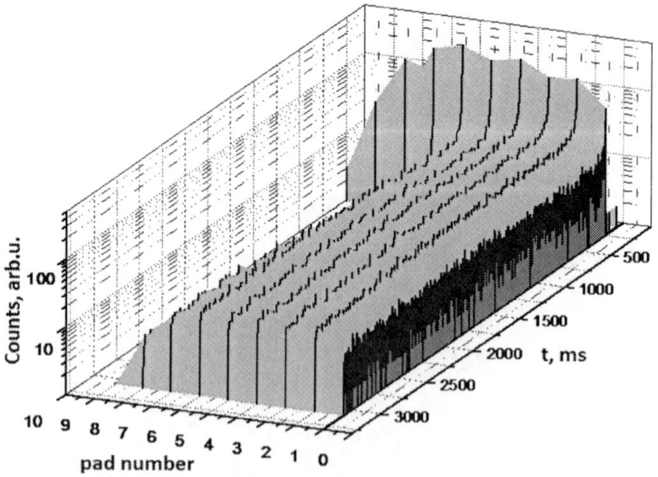

Figure 11. Example of experimental space-time ion beam structure: regime of injection without acceleration.

Conclusion

The complex for simulating film irradiation regimes has been created; this complex is a necessary part of metrological support of the industrial irradiation of polymer roll materials with the given characteristics. Numerical study of irradiation regimes elucidated the influence of such parameters as beam size, current dynamics, and film movement velocity, on the quality of produced track membranes; it was shown that to achieve film irradiation homogeneity on a level of 5–10%, the system of dynamic diagnostic of irradiation characteristics should provide measurement of beam intensity oscillations with a time resolution of order of 100 μs.

The system for dynamic diagnostics of accelerated ion beam based on transparent profilometers and noise immune charge-frequency transducers with a time resolution of 100 μs was created and installed at the industrial accelerator complex JSC "Trackpore Technology" (Dubna, Russia).

Recommendations for optimizing industrial film irradiation using this dynamic diagnostic system were developed.

The fast nondestructive highly sensitive monitoring system for circulating beams of heavy charged particles operating in a wide range of particle intensities was created based on microchannel plates and installed in the accelerator ring of the research accelerator Nuclotron at the Joint Institute for Nuclear Research (Dubna, Russia).

Charge collection system of MCP detector can be adjusted for particular experimental conditions depending on required spatial resolution and load characteristics without essential changes in the overall detector design.

The system provides monitoring of spatially resolved beam dynamics in all regimes (injection, acceleration, and extraction). This system is especially useful for monitoring of heavy ion acceleration. It should be noted that our experience in nuclear physics with internal targets demonstrated the importance of dynamic information on space-time structure of beams for such research.

References

[1] Baldin, A. A.; Baldina, E. G., *Nanotechnologies in Russia*, vol. 4, nos.9-10, 711-721 (2009).

[2] Ermolaev, A. V.; Maniakov, R. K.; Pilar, A. V., in *Proceedings of XVIII JINR International Symposium on Nuclear Electronics and Computing,* Bulgarian National Electric Company "ENERGO_1," Varna, Bulgaria, September 12–18, 2001 (Varna, 2001).

[3] Baldin, A; Feofilov, G.; Gavrilov, Yu.; Tsvinev, A.; Valiev, F., *NIM A,* Vol. 323, Issues 1-2, pp.439-444 (1992).

[4] Baldin, A. A.; Borisov, V. V.; Efimov, L .G.; Korolev, V. S.; Mikhailov, K. V.; Semashko, S. V., *JINR Rapid Comm. No.* 2 [82], pp.15-22 (1997).

[5] Valiev, F. F.; Feofilov, G. A.; Tsvinev, A. P.; Baldin, A. A.; Gavrilov, Yu. K., *JINR Rapid Comm.*, No.4(50)-91, pp.27-36 (1991).

Reviewed by G. A. Feofilov, Cand.Sci, Head of Laboratory of Ultra-High Energy Physics, V.Fock Institute for Physics, St. Petersburg State University.

In: Industrial Control Systems
Editors: R. Gilbert and A. Schultz
ISBN: 978-1-61209-988-0

Chapter 4

ADAPTIVE JACOBIAN TRAJECTORY TRACKING FOR SERIAL ROBOT MANIPULATOR PASSING THROUGH SINGULARITIES

Ali T. Hasan
Ministiry of Higher Education and Scientific Research
Baghdad, Iraq

ABSTRACT

To overcome the singularities and uncertainties in arm configurations resulting from applying the inverse kinematics model for serial robot manipulators, adaptive learning algorithm using an artificial neural network (ANN) was proposed in this chapter to solve the problem. The proposed technique does not require any prior knowledge of the kinematics model of the system being controlled, the main idea of this approach is the use of an ANN to learn the robot system characteristics rather than having to specify an explicit robot system model.

In this approach a network was trained to learn desired set of angular positions and velocities from a given set of end effector's positions and velocities. Training was performed Off-line and training data were recorded from sensors fixed on each joint.

Testing the resulting network on a general 6 DOF serial robot manipulator has shown the generality and efficiency of the proposed algorithm.

1. INTRODUCTION

The fundamental problem of robot kinematics deals with mappings between vectors in two spaces: joint space and Cartesian space. It is known that the position vector of the Cartesian space coordinates of a robot manipulator is related to the position vector of the joint coordinate by a nonlinear differentiable mapping and the velocity vector of the task space coordinates is related to that of the joint coordinate by the Jacobian matrix of the mapping. In control problems of robot manipulators, however, it is essentially important that the controller is designed to provide the transformation of both the positions and velocities from the Cartesian space to the joint space coordinate [1].

Before moving a robot arm, it is of considerable interest to know whether there are any obstacles present in its path. Computer-based robots are usually servoed in the joint space, whereas objects to be manipulated are usually expressed in the Cartesian space because it is easier to visualize the correct end-effector position in Cartesian coordinates than in joint coordinates. In order to control the position of the end-effector of the robot, an inverse kinematics (IK) solution routine should be called upon to make the necessary conversion [2], In the presence of uncertainty in kinematics, it is impossible to derive the desired joint angles from the desired Cartesian path by only solving the (IK) problem. On the other hand, the Jacobian matrix is a critical component for generating trajectories of prescribed geometry in the joint space. Most coordination algorithms employed by industrial robots avoid numerical inversion of the Jacobian matrix by driving analytical inverse solution on an ad hoc basis. Therefore, it is important that efficient algorithms be developed. It must be remarked that to achieve perfect path tracking, it is necessary to know the whole trajectory beforehand, which leads to off-line control techniques [3-5].

Robot control problem can be divided into two main areas: *kinematics control*, and *dynamic control* [6]. Handling of torque limits naturally leads to control algorithms based on the dynamic model of the manipulator as, e.g., in [7-9]. A problem with these algorithms is the remarkable computational load required to handle the dynamics of a full-sized manipulator, which is seldom affordable by current industrial control units. In addition, implementation of torque-based control laws requires replacement of the low-level joint servos typically available in industrial robots with custom control loops. As a matter of fact, to our knowledge, on-line dynamic-based methods have indeed been tested in experiments only on laboratory setup with arms of a few *degrees of freedom* (DOF).

A different approach to path tracking aimed at overcoming the above drawbacks is based on the so-called kinematics control. In detail, kinematics control consists in an inverse kinematics transformation which sends to the joint servos the reference values corresponding to an assigned end-effector trajectory (both position and velocity); as a first advantage, this allows simple interfacing with the standard control architecture of industrial robots. In the framework of kinematics-based methods for path tracking, the counterpart of the physically meaning joint torque limits is played by acceleration constraints and the use of full dynamic models can be avoided; this typically leads to computationally light algorithms that allow real-time implementation on standard numerical hardware even for robot arms of many (DOFs). A further advantage of kinematics control methods is the possibility of exploiting the presence of redundant (DOFs) [5].

Many research efforts have been devoted towards solving this problem, one of the first algorithms employed was the Resolved Motion Rate-Control method [10], which uses the pseudoinverse of the Jacobian matrix to obtain the joint velocities corresponding to a given end-effector velocity, an important drawback of this method was the singularity problem.

A velocity-singular configuration is a configuration in which a robot manipulator has lost at least one motion (DOF). In such configurations, the inverse Jacobian will not exist, and the joint velocities of the manipulator will become unacceptably large that often exceed the physical limits of joint actuators [11]. To overcome the problem of kinematics singularities, the use of a damped least squares inverse of the Jacobian matrix has been later proposed in lieu of the pseudoinverse [12,13].

Since in the above algorithmic methods the joint angles are obtained by numerical integration of the joint velocities, these and other related techniques suffer from errors due to both long-term numerical integration drift and incorrect initial joint angles. To alleviate the difficulty, algorithms based on the feedback error correction are introduced [14,15]. However, it is assumed that the exact model of manipulator Jacobian matrix of the mapping from joint coordinate to Cartesian coordinate is exactly known. It is also not sure to what extent the uncertainty could be allowed. Therefore, most research on robot control has assumed that the exact kinematics and Jacobian matrix of the manipulator from joint space to Cartesian space are known. This assumption leads to several open problems in the development of robot control laws today [5].

In recent years, there have been increasing research interest of ANNs and many efforts have been made on applications of Neural Networks to various control problems [16-20]. *Artificial neural network* (ANN) usesdatasets to obtain the models of systems in fields such as robotics, factory automation, and autonomous vehicles. Their ability to learn by example makes artificial neural networks very flexible and powerful. Therefore, neural networks have been intensively used for solving regression and classification problems in many fields. In short, neural networks are nonlinear processes that perform learning and classification. Recently neural networks have been used in many areas that require computational techniques such as pattern recognition, optical character recognition, outcome prediction and problem classification. The current focus in learning research lies on increasingly more sophisticated algorithms for the off-line analysis of finite data sets, without severe constraints on the computational complexity of the algorithms [21].

Kuroe et al. [1], have proposed a learning method of a neural network such that the network represents the relations of both the positions and velocities from the Cartesian coordinate to the joint space coordinate. They've driven a learning algorithm for arbitrary connected recurrent networks by introducing adjoint neural networks for the original neural networks (Network inversion method). On-line training has been performed for a 2 DOF robot.

In this chapter ,a learning method was propose of a neural network such that the network represents the transformation of both the positions and velocities from the Cartesian space coordinate to the joint space coordinate simultaneously, to control the motion of a 6 DOF serial robot manipulator and to overcome the arising problems, which are mainly singularities and uncertainties in arm configurations. In this approach a network was trained to learn desired set of angular positions and velocities from a given set of end effector positions and velocities. Off-line training was performed and training data were recorded from sensors fixed on each joint (as was recommended by Karlik and Aydin [6]).

2. KINEMATICSJACOBIAN OF SERIAL ROBOT MANIPULATORS

When a robot is required to perform a task in the Cartesian space coordinates, the desired Cartesian positions have to be transformed into the joint space positions. Other important information required is the velocity transformation from the Cartesian space to the joint space in order to regulate velocity at the end effector [22].

It is known that the vector of Cartesian space coordinates x of a robot manipulator is related to the joint coordinates q by:

$$x = f(q) \tag{1}$$

where $f(\cdot)$ is a nonlinear differential function.

On the other hand, the mapping from the vector of joint velocities $\dot{q}$ to the vector of Cartesian space velocities $\dot{x}$ is also of interest, which is given by:

$$\dot{x} = J(q)\dot{q} \tag{2}$$

where $J(q)$ is the Jacobianmatrix.

The inverse kinematics problem is defined as, given x the end effector position vector, q the joint variable vector is to be found. More desirably given x and $\dot{x}$ to find q and $\dot{q}$.

In other words, the inverse kinematics problem is to solve the inverse problems [1]:

$$q = f^{-1}(x) \tag{3}$$

$$\dot{q} = J^{-1}(q)\dot{x} \tag{4}$$

Solving equation (3), mapping from the joint space to the Cartesian space is uniquely decided when the end effector's position is calculated using direct kinematics, as shown in Figure 1(a). However, the transformation from the Cartesian to the joint space is not uniquely decided in the inverse kinematics as shown in Figure 1(b) [18].

On the other hand, at certain manipulator configurations, the Jacobian matrix may lose its full rank. Hence as the manipulator approaches these configurations (singular configurations), the Jacobian matrix becomes ill conditioned and may not be invertible. Under such a condition, numerical solution for equation (4) results in infinite joint rates [4]. Conventional condition for singular configuration is the determinant of the Jacobian matrix equal to zero, namely *det (J)=0.*

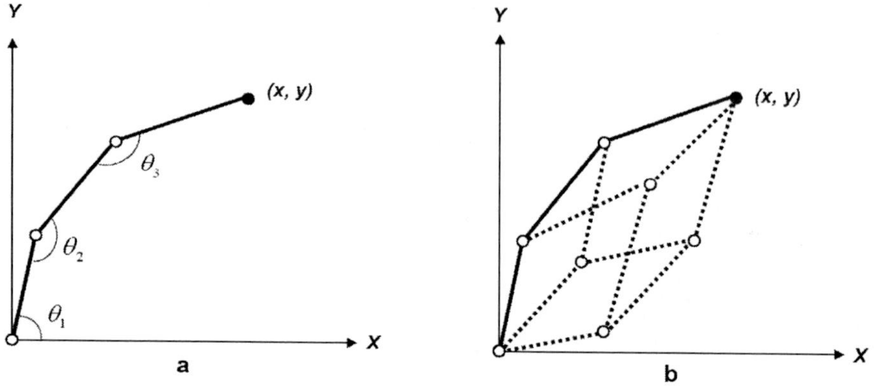

Figure 1.Three DOF robot arm. a) Joint angles and end-effector's coordinates (forward kinematics). b) Combination of all possible joint angles (Inverse Kinematics).

To analyze the singular conditions and to develop effective algorithms to resolve the inverse kinematics problem at or in the vicinity of singularities are of great importance.

3. Artificial Neural Networks (ANNs)

An ANN consists of massively interconnected processing nodes known as neurons. Each of which accepts a weighted set of inputs and responds with an output. (Knowledge acquired by the network is stored as a set of connection weights).

Sum of the weighted inputs is processed through an activation function; basically, the neuron model represents the biological neuron that fires when its inputs are significantly excited. There are many ways to define the activation function such as threshold function, sigmoid function, and hyperbolic tangent function.

An ANN can be trained to perform a particular function by adjusting the values of connections, i.e., weighting coefficients, between the processing nodes. In general, ANNs are adjusted / trained to reach from a particular input to a specific target output using a suitable learning method until the network output matches the target. The error between the output of the network and the desired output is minimized by modifying the weights. When the error falls below a determined value or the maximum number of epochs is exceeded, training process is ceased. Then, this trained network can be used for simulating the system outputs for the inputs that have not been introduced to the network before.

The architecture of an ANN is usually divided into three parts: an input layer, a hidden layer(s) and an output layer. The information contained in the input layer is mapped to the output layer through the hidden layer(s). Each unit can send its output to the units only on the higher layer and receive its input from the lower layer. For a given modelling problem, the numbers of nodes in the input and output layers are determined from the physics of the problem, and equal to the numbers of input and output parameters respectively, while the number of nodes in the hidden layers(s) is determined by trial and error [21].

4. Collecting Training Data

Different researchers have applied different methods for gathering training data, some of them have used the kinematics equations [6,21], some of them have used the network inversion method [1,18], some of them have used the cubic trajectory planning [23] and other some have used a simulation program for this purpose [24].

However, there are always kinematics uncertainties presence in the real world such as ill-defined linkage parameters, links flexibility and backlashes in gear train, in this approach, training data were recorded directly from sensors fixed on each joint (as was recommended by Karilk and Aydin[6]).

5. Network's Implementation

A supervised feed forward ANN was designed using *C programming language* to overcome the uncertainties in the arm configurations and the singular configurations. The network consists of input, output and one hidden layer, the input vector for the network consists of the position of the end effector of the robot along the X, Y and Z coordinates of the global coordinate system and the end-effector's linear velocity, while the output vector was the angular position and velocity of each of the 6 joints as can be seen in Figure 2, every neuron in the network is fully connected with each other, sigmoid transfer function was chosen to be the activation function, generalized backpropagation delta learning rule (GDR) algorithm was used in the training process.

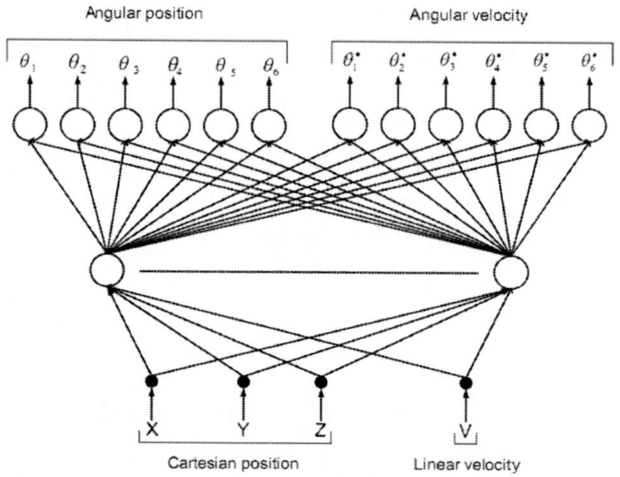

Figure 2.The topology of the designed artificial neural network.

600 data sets were recorded experimentally from sensors fixed on the FANUC M-710i robot joints which is a 6 DOF robot, 400 set were used for the training while the other 200 sets were used for the testing process.

All input and output values are scaled individually such that overall variance in data set is maximized, this is necessary as it leads to faster learning, all the vectors were scaled to reflect continuous values ranges from –1 to 1.

5.1. Training Stage

Determining right number of hidden neurons is one of the most important considerations when training any network; the problem of the number choice is under intensive study with no conclusive answers available. Getting the right number of hidden neurons is a matter of trial and error, since there is no science to it.

Number of hidden neurons was set to be 77 with a learning factor of 0.9 by trial and error.

The difference between desired and actual system output could be the performance measure for the network, as a learning system adapts its internal structure to achieve better response.

In the GDR the system is modified following each iteration, which leads to the learning curve shown in Figure 3.

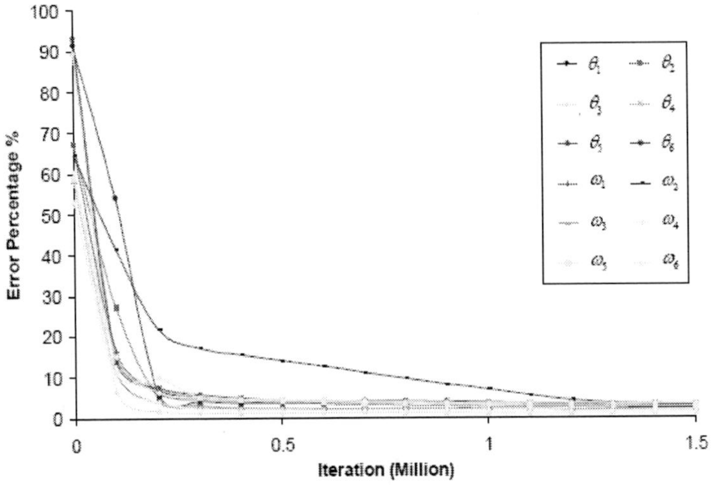

Figure 3.Building knowledge curve for the training data set.

Table 1 shows the error percentages of each of the six joints after the training was finished.

Table 1. Total error percentages for the training data set

	Joint 1	Joint 2	Joint 3	Joint 4	Joint 5	Joint 6
Angular Position θ	2.898%	1.673%	0.887%	3.145%	3.152%	2.175%
Angular Velocity ω	2.665%	3.115%	1.688%	3.27%	2.563%	1.625%

5.2. Testing Stage

To test the ability of the trained network to make prediction for a new path which has never been introduced to the network before, and as evidence that the designed network is capable of making well generalization later overcoming the uncertainty and singularity problem arising from using the robot model.

The trained network was tested for a new trajectory, which has never been seen by the network before, the testing data were meant to pass nearby and through the singular configurations (Fourth and Fifth joints), these configurations have been determined by setting the determinant of the Jacobian matrix to zero.

Table 2 shows the percentages of error in the testing data set that consists of 200 points, while Figures 4 and 5 show the trajectory tracking of the testing data of the joints where singular configurations are taking place.

Table 2. Total error percentages for the testing data set

	Joint 1	Joint 2	Joint 3	Joint 4	Joint 5	Joint 6
Angular position θ	1.8%	0.245%	0.2%	5.32%	9.885%	0.08%
Angular Velocity ω	3.82%	2.875%	1.47%	2.64%	3.28%	1.41%

A good trajectory tracking can be seen through these figures even when the robot is passing nearby and through the singular configurations.

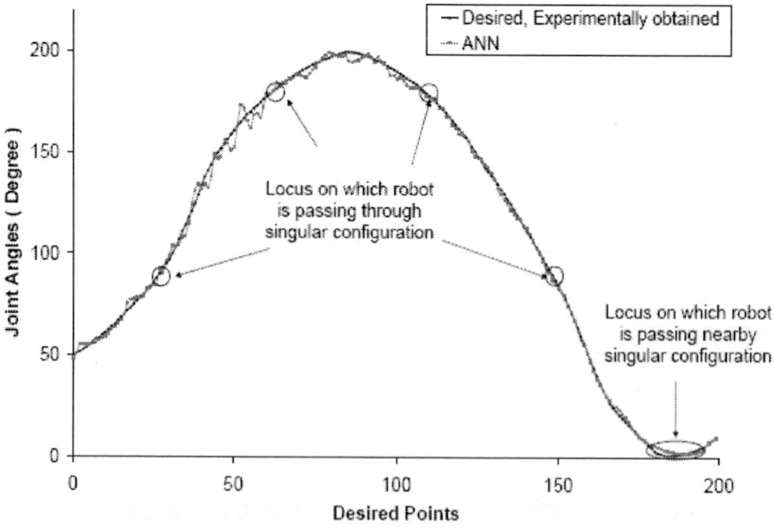

Figure 4.Predicted angular position for joint 4.

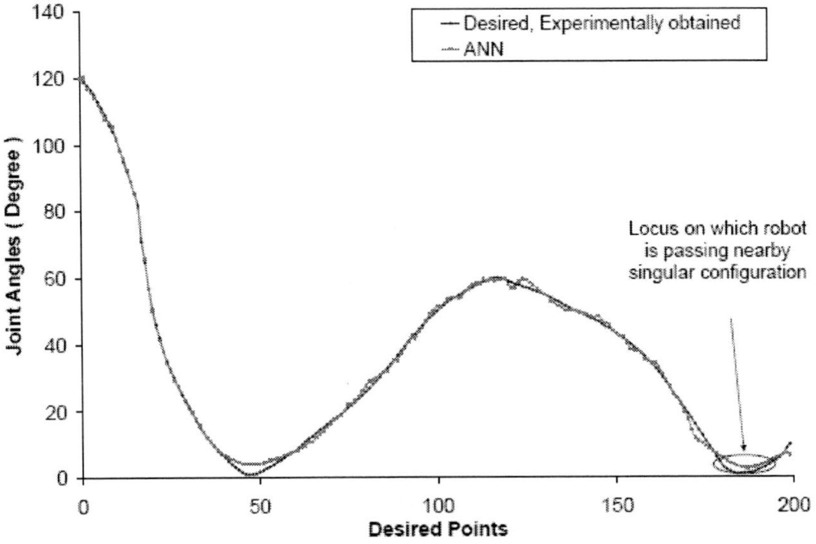

Figure 5.Predicted angular position for joint 5.

CONCLUSIONS

In order to overcome the drawbacks of some control schemes which depends onmodeling the system being controlled, ANN technique has been utilized where learning is done iteratively based only on observation of input-output relationship unlike most other control schemes, which is a significant advantage of using ANN technology. The aim of using neural network is to find a set of weights that ensure that for each input vector the output vector produced by the network is the same as or sufficiently close to the desired output vector, so, system model does not have to be known at the time of the controller design, also, any change in the physical setup of the system suchas the addition of a new tool would only involve training and will not require any major system software modifications.

This approach possesses several distinct advantages: First, it can be applied to any general serial manipulator with positional degrees of freedom since learning is only based on observations of input / output relationships of the system being controlled.

Second, high accuracy can be achieved along the desired path. Third, if available, the proposed approach can be adapted to any general serial manipulator including both redundant and non-redundant systems.

For future research, another configuration (using another learning mode, another activation function, or another topology) can be tested in order to reach, if possible, a more accurate path.

REFERENCES

[1] Kuroe, Y., Nakai, Y. and Mori, T., 1994. A new Neural Network Learning on Inverse Kinematics of Robot Manipulators. International Conference on Neural Networks, *IEEE world congress on computational Intelligence.* Vol.5: 2819-2824.

[2] Fu, K.S., Gonzalez, R.C. and Lee, C.S.G., 1987. Robotics Control, Vision, and Intelligence. McGraw-Hill book Co. Singapore, international edition.

[3] Cheah, C.C., Lee, K., Kawamura, S., and Arimoto, S., Asymptotic stability of robot control with approximate Jacobian matrix and its application to visual servoing. *Proceeding of the 39th IEEE conference on decision and control,* Sydney, Australia: 3939-3944.2000.

[4] Tsai, L.W., 1999. *Robot analysis, the mechanics of serial and parallel manipulators.* John Willy and Sons, Inc.

[5] Antonelli, G., Chiaverini, S. and Fusco, G., 2003. A new on-line algorithm for inverse kinematics of robot manipulators ensuring path-tracking capability under joint limits. *IEEE Transaction on Robotics and Automation,* Vol.19, No.1: 162-167.

[6] Karilk, B., Aydin, S., 2000. An improved approach to the solution of inverse kinematics problems for robot manipulators. *Engineering applications of artificial intelligence*. Vol.13: 159-164.

[7] Dahl, O. and Nielsen, L., 1990. Torque-limited path following by on line trajectory time scaling. *IEEE Transaction on Robotics and Automation*, vol. 6: 554–561.

[8] Kirc′anski , M. and Kirc′anski, N., 1997. Resolved-rate and resolved-acceleration based robot control in the presence of actuators constraints, In Proc. *1997 IEEE Int. Conf. Robotics and Automation, Albuquerque*, NM: 235–240.

[9] Faiz, N. and Agrawal, S. K., 2000. Trajectory planning of robots with dynamics and inequalities. *In Proc. IEEE Int. Conf. Robotics and Automation*, San Francisco, CA: 3977–3983.

[10] Whitney. E., 1969. Resolved motion rate control of manipulators and human prostheses, *IEEETransaction Man–Mach. Systems*, vol. MMS–10:47–53.

[11] Hu, Z., FU, Z. and Fang, H.,2002. Study of singularity robust inverse of Jacobian matrix for manipulator. *Proceedings of the First International Conference on Machine Learning and Cybernetics,* Beijing: 406-410.

[12] Nakamura, Y. and Hanafusa, H., 1986. Inverse kinematic solutions with singularity robustness for robot manipulator control. *Journal of Dynamic Systems Measurements Control*, Vol. 108: 163–171.

[13] Wampler, C. W., 1986. Manipulator inverse kinematic solutions based on vector formulations and damped least-squares methods. *IEEE Transaction Syst., Man, Cybernetics.* Vol. 16: 93–101.

[14] Balestrino, A., De Maria, G. and Sciavicco, L., 1984. Robust control of robotic manipulators. *International Proceedings of the9th IFAC World Congr.,* Vol. 6, Budapest, Hungary:80-85.

[15] Wampler, C. W. and Leifer, L. J., 1988. Applications of damped least-squares methods to resolved-rate and resolved-acceleration control of manipulators, *Journal of Dynamic Systems Measurements Control*, Vol. 110: 31–38.

[16] D'Souza, A.,Vijayakumar, S., and Schaal, S., 2001. Learning Inverse Kinematics.*Proceedings of the 2001 IEEE/ RSJ International Conference on Intelligent Robots and Systems Maui*, Haw- USA: 298-303.

[17] Ogawa, T., Matsuura, H., and Kanada, H., A Solution of Inverse Kinematics of Robot Arm Using Network Inversion.*Proceedings of the 2005 International Conference on Computational Intelligence for Modelling, Control and Automation.*

[18] Köker, R., 2005. Reliability-based approach to the inverse kinematics solution of robots using Elman's networks. *Engineering Applications of Artificial Inttelegence.*Vol.18: 685-693.

[19] Hasan, A.T., Hamouda, A.M.S., Ismail, N., and Al-Assadi, H.M.A.A., 2006. An adaptive-learning algorithm to solve the inverse kinematics problem of a 6 D.O.F serial robot manipulator. *Advances in Engineering Software.*Vol.37: 432-438.

[20] Al-Assadi, H.M.A.A., Hamouda, A.M.S., Ismail, N. and Aris, I., 2007. An adaptive learning algorithm for controlling a two-degree-of-freedom serial ball-and-socket actuator .*Proceedings of the I MECH E Part I Journal of Systems and Control Engineering,* Vol.221, No. 7 :1001-1006.

[21] Bingual, Z., Ertunc, H.M. and Oysu, C., Comparison of Inverse Kinematics Solutions Using Neural Network for 6R Robot Manipulator with Offset. *2005 ICSC congress on Computational Intelligence*.

[22] Jung, S. and Ravani, B., 1998. On-line Kinematics Jacobian Uncertainty Compensation for Robot Manipulators Using Neural Network.*IEEEInt.Conf. on system, Man, and cybernetics*.Vol.4: 3483-3488.

[23] Köker, R.,Öz, C.,Çakar.T. andEkiz, H., 2004. A study of neural network based inverse kinematics solution for a three-joint robot. *Robotics and Autonomous Systems*.Vol. 49: 227–234.

[24] Driscoll, J.A., Comparison of neural network architectures for the modeling of robot inverse kinematics. *Proceedings of the IEEE, south astcon*.2000:44-51.

In: Industrial Control Systems
Editors: R. Gilbert and A. Schultz
ISBN: 978-1-61209-988-0

Chapter 5

INTELLIGENT CONTROL SYSTEM FOR AN INDUSTRIAL MANIPULATOR

Fusaomi Nagata
Tokyo University of Science, Yamaguchi, Japan
Keigo Watanabe
Okayama University, Japan

Abstract

In this chapter, two types of intelligent control systems are introduced. One is the adaptive learning technique for large-scale teaching signals. Recently various control methods as represented by PID control are used for robotic control. To cope with the requirements for high response and precision, advanced feedforward controllers such as gravity compensator, Coriolis/centrifugal force compensator and friction compensators have been built in the controller. Generally, it causes heavy computational load when calculating the compensating value within a short sampling period. To overcome the problem, integrated recurrent neural networks are applied as a feedforward controller for PUMA560 manipulator. The feedforward controller works instead of gravity and Coriolis/centrifugal force compensators. In the learning process of neural network by using back propagation algorithm, in general, the learning coefficient and gain of sigmoid function are tuned intuitively and empirically according to teaching signals. The tuning is complicated because of being conducted by trial and error. Especially, when the scale of teaching signal is large, the problem becomes crucial. To cope with the problem concerning the learning performance, a simple and adaptive learning technique for large-scale teaching signals is proposed. The learning techniques and control effectiveness are evaluated through simulations using the dynamic model of PUMA560 manipulator.

The other is the fine gain tuning technique for model-based robotic servo controllers using genetic algorithms. Resolved acceleration control method and computed torque control method are used for nonlinear control of industrial manipulators, which are composed of a model base portion and a servo portion. The servo portion is a closed-loop system with respect to the position and velocity. On the other hand, the model base portion has the inertia term, gravity term and centrifugal/Coriolis term, which works for canceling the nonlinearity of manipulator. In order to realize high control stability, the position and velocity feedback gains used in the servo portion should be tuned suitably. In the latter half of the chapter, a simple but effective fine tuning method after the manual tuning process is introduced for the position and velocity feedback gains in the servo portion. At the first step, the search space for the gains are roughly narrowed down by a controller designer, e.g., considering the critically damped condition. At the next step, the gains are finely tuned within the space by using genetic algorithms. The genetic algorithms search for the better combination of the position and velocity gains. Simulations are conducted using the dynamic model of PUMA560 manipulator to validate the effectiveness of the proposed method.

Keywords: Neural networks, Large-scale teaching signal, Sigmoid function, Adaptive learning, Servo system, PUMA560 manipulator, Trajectory following control, Nonlinear control, Model based servo controller, Fine gain tuning, Genetic algorithms, Critically damped condition

1 Introduction

In this chapter, two types of intelligent control systems are introduced. One is the adaptive learning technique for large-scale teaching signals. Recently various control methods as represented by PID control are used for robotic control. To cope with the requirements for high response and precision, advanced feedforward controllers such as gravity compensator, Coriolis/centrifugal force compensator and friction compensators have been built in the controller. Generally, it causes heavy computational load when calculating the compensating value within a short sampling period. To overcome the problem, integrated recurrent neural networks are applied as a feedforward controller for a PUMA560 manipulator. The feedforward controller works instead of gravity and Coriolis/centrifugal force compensators. In the learning process of neural network by using back propagation algorithm, in general, the learning coefficient and gain of sigmoid function are tuned intuitively and empirically according to teaching signals. The tuning is complicated because of being conducted by trial and error. Especially, when the scale of teaching signal is large, the problem becomes crucial. To cope with the problem concerning the learning performance, a simple and adaptive learning technique for large-scale teaching signals is proposed. The learning techniques and control effectiveness are evaluated through simulations using the dynamic model of a PUMA560 manipulator.

The other is the fine gain tuning technique for model-based robotic servo controllers using genetic algorithms. Resolved acceleration control method and computed torque control method are used for nonlinear control of industrial manipulators, which are composed of a model base portion and a servo portion. The servo portion is a closed-loop system with respect to the position and velocity. On the other hand, the model base portion has the inertia term, gravity term and centrifugal/Coriolis term, which works for canceling the nonlinearity of manipulator. In order to realize high control stability, the position and velocity feedback gains used in the servo portion should be tuned suitably. In the latter half of the chapter, a simple but effective fine tuning method after the manual tuning process is introduced for the position and velocity feedback gains in the servo portion. At the first step, the search space for the gains are roughly narrowed down by a controller designer, e.g., considering the critically damped condition. At the next step, the gains are finely tuned within the space by using genetic algorithms. The genetic algorithms search for the better combination of the position and velocity gains. Simulations are conducted using the dynamic model of the PUMA560 manipulator to validate the effectiveness of the proposed method.

2 Adaptive Learning Technique for Large-Scale Teaching Signals

2.1 Background

In this decade, open architecture industrial robots have been produced from several industrial robot makers such as KAWASAKI Heavy Industries, Ltd., MITSUBISHI Heavy Industries, Ltd. and YASKAWA Electric Corp. and so on. The open architecture described in this article means that the servo system and the kinematics of the robot are technically opened, so that various applications required in industrial fields are allowed to be planned and developed at the user side. For example, non-taught operation by using a CAD/CAM system can be considered due to the technically opened accurate kinematics. Also, force control strategy using a force sensor can be easily implemented on the technically opened discrete-time servo system [?].

It is now possible to model and simulate many types of robots. For example, Chen et al. presented a new design of an environment for simulation, animation, and visualization of sensor-driven robots. Although conventional computer-graphics-based robot simulation and animation software packages lacked the capabilities for robotic sensing simulation, the system was designed to overcome the deficiency [?]. Also, Benimeli et al. addressed the implementation and comparison of an indirect and a direct identification procedures on an industrial robot provided with an open control architecture. The estimation of dynamic parameters in mechanical systems constituted an issue of crucial importance for dynamic simulations where high accuracy was required [?].

We have introduced a simulation technique of velocity-based discrete-time control system for open architectural industrial robots [?]. In order to develop a novel velocity-based control system represented in discrete-time domain for an open architecture industrial robot, it is required from the viewpoint of safety, cost and easiness to preliminarily examine and evaluate the characteristics and performance. In such a case, the proposed simulation techniques are useful.

Neural networks, which are one of the typical intelligent control approaches, are paid attention to improving the control performance of robotic system. Han and Moraga reported a fundamental and important result in which a sigmoid function was employed in each unit. A variant sigmoid function with three parameters that denoted the dynamic range, symmetry and slope of the function was discussed. How these parameters influenced the speed of back propagation learning was illustrated, and a hybrid sigmoidal network with different parameter configuration in different layers was introduced. The error signal problem, oscillation problem and asymmetrical input problem could be handled by regulating and modifying the sigmoid function parameter configuration in different layers [?].

As for the application of neural networks to robotic control, many control systems and their learning techniques have been developed. For example, Torras described that to carry out their tasks autonomously in unknown environments, it is essential for robots to adapt to their environments. The most effective method to endow robots with this capability is through the use of neural networks. The related techniques were explained with mobile robots and manipulator arms [?]. Hasan et al. proposed an adoptive learning strategy using an artificial neural network to control the motion of a manipulator robot with six degree-of-freedom and to overcome the inverse kinematics problem, which are mainly singularities and uncertainties in arm configurations [?]. Horng showed that LMBP (Levenberg-Marquardt back-propagation) has faster convergence compared with other three back-propagation modified algorithms and it is suitable for system identification and controller design. MATLAB/Simulink and LabVIEW with neural networks were successfully integrated to develop a SCADA (Supervisory Control AND Data Acquisition) system of AC servo motor [?]. Also, Huang et al. developed an adaptive neural network algorithm for a class of interconnected nonlinear systems. Neural networks were applied to approximate the unknown nonlinear functions and interconnections in the subsystems. A systematic approach was established to synthesize the adaptive learning control scheme. The effectiveness was demonstrated by simulations [?]. Up to now, however, it seems that the consideration about the scale of teaching signal in the output layer has not been sufficiently taken yet.

Recently various control methods as represented by PID control are used for robotic control. To cope with the requirements for high response and precision, advanced feedforward controllers such as gravity compensator, Coriolis/centrifugal force compensator and friction compensators have been built in the controller. However, it generally requires a serious computational effort to calculate the compensating value within a short sampling period.

In the first half of the chapter, a recurrent neural network is applied for a feedforward controller of the PUMA560 manipulator which is one of typical industrial robots with six degree-of-freedom. Hereafter, the recurrent neural network is called RNN. The feedforward controller includes gravity compensator and Coriolis/centrifugal force one. Also, a simple and adaptive learning technique is proposed for a robotic servo system to deal with large-scale joint driving torques [?]. The learning technique and control effectiveness are evaluated through simulations using the dynamic model of PUMA560 manipulator as shown in Fig. ??. The subsections are organized as follows: Subsection 2.2 explains the computed torque control method used as a model-based servo system. Then, teaching signals for RNNs are calculated. In subsection 2.3, integrated RNNs are designed to compensate nonlinear terms such as gravity term and Coriolis/centrifugal force term. Further, a simple and adaptive learning technique by using a scaler is proposed for large-scale teaching signals processed in back propagation algorithm, then the effectiveness is discussed. In subsection 2.4, a servo system using the integrated RNNs is applied to a trajectory following control problem. The result demonstrates that the integrated RNNs can work well instead of gravity and Coriolis/centrifugal force terms.

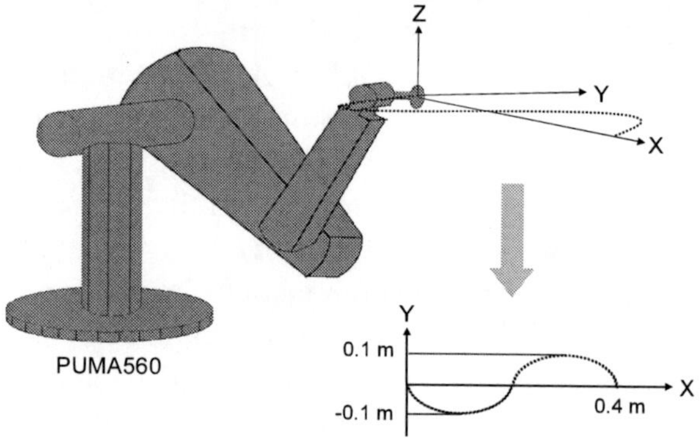

Figure 1. Image of a desired trajectory composed of a sine curve.

2.2 Model-based robotic servo system

In order to simulate an industrial robot, first of all, a servo system is considered and designed. Here, the computed torque control method [?, ?] is used in a servo system. The computed torque control method is used for nonlinear control of industrial manipulators, which is composed of a model base portion and a servo portion. The servo portion is a closed loop with respect to the position and velocity. On the other hand, the model base portion has the inertia term, gravity term and centrifugal/Coriolis term, which work for canceling the nonlinearity of manipulator. In order to realize high control stability, the position and velocity feedback gains used in the servo portion should be tuned suitably.

A simple desired trajectory as shown in Fig. **??** is calculated in advance for the computed torque control method. The computed torque control method generates joint driving torques to follow the trajectory. The desired trajectory and joint driving torque are used as teaching signals for neural networks.

2.2.1 Computed torque control method

The dynamic model of a manipulator with the friction term is generally given by

$$\boldsymbol{M}(\boldsymbol{\theta})\ddot{\boldsymbol{\theta}} + \boldsymbol{H}(\boldsymbol{\theta}, \dot{\boldsymbol{\theta}}) + \boldsymbol{G}(\boldsymbol{\theta}) + \boldsymbol{F}_r(\boldsymbol{\theta}, \dot{\boldsymbol{\theta}}) = \boldsymbol{\tau} \tag{1}$$

where, $\boldsymbol{M}(\boldsymbol{\theta}) \in \Re^{6\times6}$ is the inertia term in joint space. $\boldsymbol{H}(\boldsymbol{\theta}, \dot{\boldsymbol{\theta}}) \in \Re^{6\times1}$ and $\boldsymbol{G}(\boldsymbol{\theta}) \in \Re^{6\times1}$ are the Coriolis/centrifugal force term and gravity term, respectively. $\boldsymbol{F}_r(\boldsymbol{\theta}, \dot{\boldsymbol{\theta}}) \in \Re^{6\times1}$ is the friction term consisting of viscous friction and Coulomb friction. $\boldsymbol{\theta} \in \Re^{6\times1}$, $\dot{\boldsymbol{\theta}} \in \Re^{6\times1}$ and $\ddot{\boldsymbol{\theta}} \in \Re^{6\times1}$ are the angle, its velocity and acceleration vectors in joint coordinate system, respectively. $\boldsymbol{\tau} \in \Re^{6\times1}$ is the joint driving torque vector. In the case that the computed torque control law is employed in the servo system of a manipulator, the desired angle, angle velocity and acceleration vectors in joint coordinate system are respectively given to the references of the servo system, so that the joint driving torque can be calculated from

$$\boldsymbol{\tau} = \hat{\boldsymbol{M}}(\boldsymbol{\theta}) \left[\ddot{\boldsymbol{\theta}}_r + \boldsymbol{K}_v\{\dot{\boldsymbol{\theta}}_r - \dot{\boldsymbol{\theta}}\} + \boldsymbol{K}_p\{\boldsymbol{\theta}_r - \boldsymbol{\theta}\}\right] + \hat{\boldsymbol{H}}(\boldsymbol{\theta}, \dot{\boldsymbol{\theta}}) + \hat{\boldsymbol{G}}(\boldsymbol{\theta}) \quad (2)$$

where, $\hat{}$ denotes the modeled term. $\boldsymbol{\theta}_r \in \Re^{6\times1}$, $\dot{\boldsymbol{\theta}}_r \in \Re^{6\times1}$ and $\ddot{\boldsymbol{\theta}}_r \in \Re^{6\times1}$ are the desired angle, angle velocity and acceleration vectors. $\boldsymbol{K}_v = \mathrm{diag}(K_{v1}, \ldots, K_{v6})$ and $\boldsymbol{K}_p = \mathrm{diag}(K_{p1}, \ldots, K_{p6})$ are the feedback gains of velocity and position, respectively. Note that $\boldsymbol{\theta}$, $\dot{\boldsymbol{\theta}}$ in Eq. (**??**) are actual values, i.e., controlled variables. The nonlinear compensation terms $\hat{\boldsymbol{H}}(\boldsymbol{\theta}, \dot{\boldsymbol{\theta}})$ and $\hat{\boldsymbol{G}}(\boldsymbol{\theta})$ are calculated to cancel the nonlinearity and are effective to achieve stable trajectory control. The block diagram of the computed torque control method is shown in Fig. **??**. Although the gravity term and centrifugal/Coriolis force term are ordinarily obtained by computing the inverse dynamics through recursive Newton-Euler formulation, the computational load within a sampling period is considerably large.

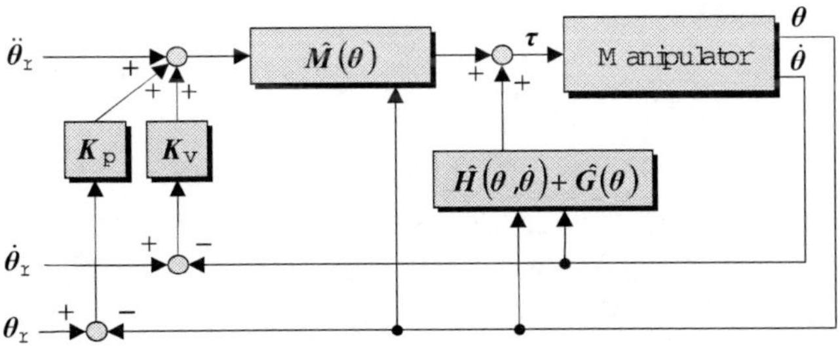

Figure 2. Block diagram of the computed torque control method, where $\boldsymbol{\theta}_r$, $\dot{\boldsymbol{\theta}}_r$ and $\ddot{\boldsymbol{\theta}}_r$ are the desired angle, velocity and acceleration vectors in joint coordinate system.

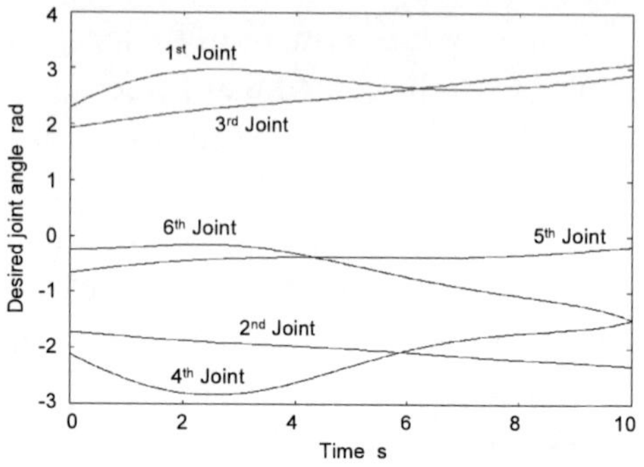

Figure 3. Desired joint angle vector $\boldsymbol{\theta} \in \Re^{6\times 1}$ realizing the trajectory shown in Fig. ??.

2.2.2 Teaching signal for RNN

As an example, the desired trajectory in Cartesian space has been designed as shown in Fig. ??. The tip of the robot arm draws a sine curve in x-y plane. The trajectory is resolved into ones in joint coordinate system by using the inverse kinematics of PUMA560, which are shown in Fig. ??. Also, Fig. ?? shows the joint driving torques needed for realizing the trajectory shown in Fig. ??, which are obtained through a trajectory following simulation by using Eq. (??). The simulation was carried out by using the dynamic model of PUMA560 manipulator on MATLAB system [?, ?]. The robotic dynamic model with the friction torque term given by Eq. (??) was applied. The friction torque term $\boldsymbol{F}_r(\boldsymbol{\theta}, \dot{\boldsymbol{\theta}})$ consists of the viscous friction torque and Coulomb friction torque, which is represented by

$$\boldsymbol{F}_r(\boldsymbol{\theta}, \dot{\boldsymbol{\theta}}) = \boldsymbol{B}\boldsymbol{G}_r^2\dot{\boldsymbol{\theta}} + \boldsymbol{G}_r\boldsymbol{\tau}_c\{\mathrm{sign}(\dot{\boldsymbol{\theta}})\} \tag{3}$$

where $\boldsymbol{B}$ is the coefficient matrix of viscous friction at each motor, $\boldsymbol{G}_r$ is the reduction gear ratio matrix which represents the motor speed to joint speed, and $\boldsymbol{\tau}_c\{\mathrm{sign}(\dot{\boldsymbol{\theta}})\}$ is the Coulomb friction torque appeared at each motor. If $\mathrm{sign}(\dot{\theta}_i) > 0$, then $\tau_{ci} = \tau_{ci}^+$; if $\mathrm{sign}(\dot{\theta}_i) < 0$, then $\tau_{ci} = \tau_{ci}^-$. In simulations, $\boldsymbol{B}$ and $\boldsymbol{G}_r$ are set to diag(0.0015, 0.0008, 0.0014, 0.0001, 0.0001, 0.0000) and diag(-62.6, 107.8, -53.7, 76.0, 71.9, 76.7), respectively; $\boldsymbol{\tau}_c^+$ and $\boldsymbol{\tau}_c^-$ are set to $[0.395\ 0.126\ 0.132\ 0.011\ 0.009\ 0.004]^T$ and $[-0.435\ -0.071\ -0.105\ -0.017\ -0.015\ -0.011]^T$, respectively [?].

After the section, six RNNs are respectively tried to acquire the teaching signals composed of desired joint angle and joint driving torque.

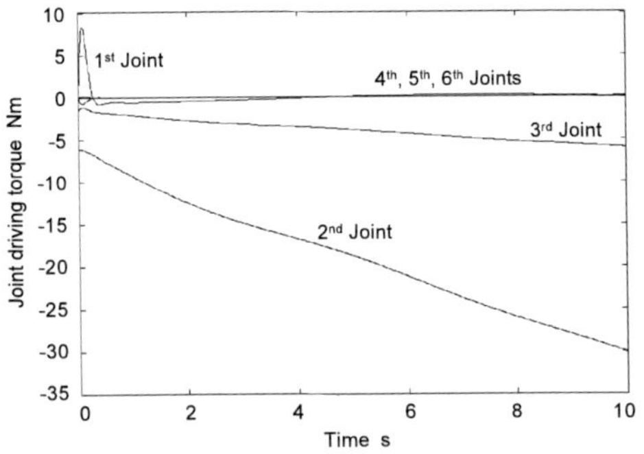

Figure 4. Desired joint driving torque vector $\boldsymbol{\tau} \in \Re^{6\times 1}$ performing the trajectory shown in Fig. **??**.

2.3 Independent recurrent neural networks for an industrial robot with six joints

2.3.1 Adaptive learning of RNNs

In order to learn the input/output relation shown in Figs. **??** and **??**, a RNN is designed as shown in Fig. **??**. Each unit has a sigmoid function as shown in Fig. **??**. The RNN consists of an input layer, two hidden layers and an output layer. Each layer has seven, thirty and one units, respectively. Note that the output layer does not have an activation function, i.e., sigmoid function. Therefore, the RNN directly yields a joint driving torque from the calculation of weighted sum. The RNN acquires the relation between the desired trajectory in joint space and joint driving torque. Although the improvement of learning performance in employing such a sigmoid function is very important, it seems that the adaptability for large-scale teaching signals has not been discussed clearly.

In this subsection, the learning process of RNN is explained in detail. The input to the hidden layers and output layer is a weighted sum of the previous layer. The sum is squashed into ± 0.5 by the sigmoid function, which is written by

$$f(X) = \frac{1}{1+\exp(-X)} - 0.5 \tag{4}$$

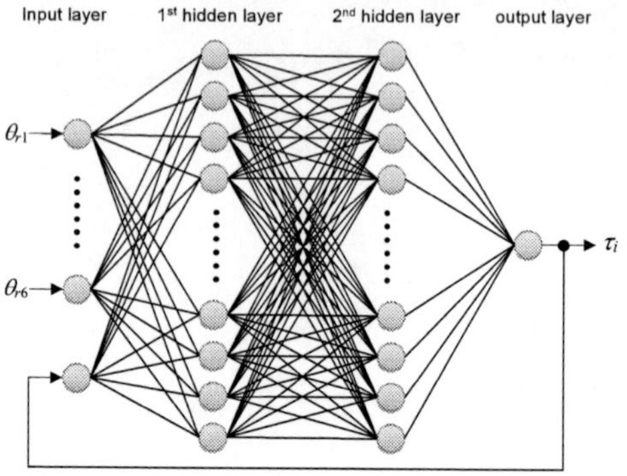

Figure 5. Recurrent type neural network to learn the input/output relation between six joint angles θ_{ri} and driving torque τ_i $(i = 1, \ldots, 6)$.

The input/output relation of each unit is given by

$$X_{i,l}(k) = \sum_{j=1}^{m} w_{j,l-1}^{i,l}(k)\, o_{j,l-1}(k) \tag{5}$$

where $X_{i,l}(k)$ is the state of i-th unit in l-th layer when k-th teaching signal is applied. $w_{j,l-1}^{i,l}(k)$ is the interconnection weight between the i-th unit in l-th layer and the j-th unit in $(l-1)$th layer. $o_{j,l-1}(k)$ is the output of the j-th unit in $(l-1)$th layer. m is the number of units in $(l-1)$th layer. Back propagation algorithm is employed in the learning process. The learning was iterated until error function E became small sufficiently. For example, in the case of first joint, the error function E is calculated by

$$E = \frac{1}{2}\sum_{k=1}^{1000} \{S_1\tau_1^*(k) - \tau_1(k)\}^2 \tag{6}$$

where $\tau_1^*(k)$ and $\tau_1(k)$ are the teaching signal and output of RNN at the discrete time $k(1 \le k \le 1000)$ with respect to the first joint, respectively. That means the pattern number of teaching signals is 1000. Also, S_1 is an adaptive element called the down scaler, which is automatically extracted from the teaching

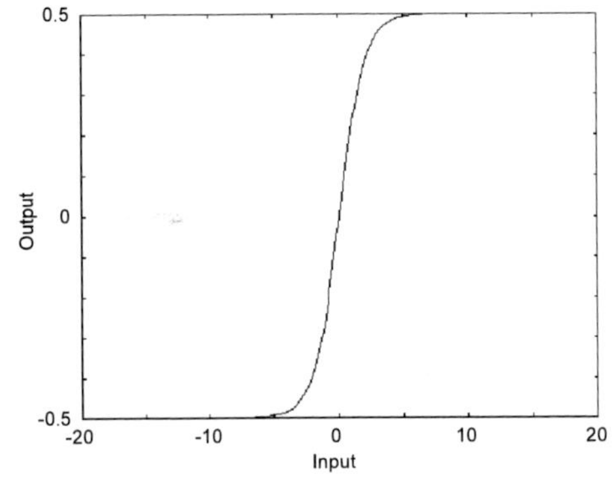

Figure 6. Sigmoid function used in each unit allocated in input and hidden layers.

signals in output layer as

$$S_1 = \begin{cases} 1 & \text{if } \max(|\tau_1(k)|) \leq \lambda \\ \dfrac{\lambda}{\max(|\tau_1(k)|)} & \text{otherwise} \end{cases} \tag{7}$$

where $\max(|\tau_1(k)|)$ is the maximum absolute value in teaching signals. λ is determined by considering the activation area in which $f(X)$ is steeply varying from about -0.5 to 0.5 as shown in Fig. **??**. For example, when the gain of sigmoid function is 1 as given by Eq. (**??**), λ can be set to about 5.

$w_{j,l-1}^{i,l}(k)$ is updated by the steepest descent method given by

$$w_{j,l-1}^{i,l}(k+1) = w_{j,l-1}^{i,l}(k) - \eta \frac{\partial E}{\partial w_{j,l-1}^{i,l}(k)} \tag{8}$$

where η is the learning coefficient. If l-th layer is the output one, i.e. $l = 4$, the update quantity for a weight is calculated by

$$\frac{\partial E}{\partial w_{j,3}^{1,4}(k)} = -\delta_{1,4}\, o_{j,3}(k) \tag{9}$$

In the above equation,

$$\delta_{1,4} = \{S_1 \tau_1^*(k) - \tau_1(k)\} f'\{X_{1,4}(k)\} \tag{10}$$

Also, in case of the third layer, the update quantity is computed by

$$\frac{\partial E}{\partial w_{j,2}^{i,3}(k)} = -\delta_{i,3}\, o_{j,2}(k) \tag{11}$$

In the above equation,

$$\delta_{i,3} = \delta_{1,4}\, w_{i,3}^{1,4}(k)\, f'\{X_{i,3}(k)\} \tag{12}$$

In the same manner, $w_{j,1}^{i,2}(k)$ is updated through

$$\frac{\partial E}{\partial w_{j,1}^{i,2}(k)} = -\delta_{i,2}\, o_{j,1}(k) \tag{13}$$

where

$$\delta_{i,2} = \sum_{n=1}^{30} \delta_{n,3}\, w_{i,2}^{n,3}(k)\, f'\{X_{i,2}(k)\} \tag{14}$$

After sufficient learning, it is expected that the RNN will give an output similar to the torque curve of the first joint as shown in Fig. **??**. Other five RNNs designed for the joints from the second to the sixth can be also learned in the same manner.

If an unknown trajectory is given and the dynamic torque for teaching signal is out of the nonlinear range, then the torque is adaptively squashed into the range within about ± 5 by using Eqs. (**??**) and (**??**) in the learning process.

2.3.2 Learning results of RNNs

In this subsection, the learning process of RNN is described. If orders of six torques in Fig. **??** were almost same, then the output layer could have six units. However, for example, the torque from -30 to -5 Nm is required for the second joint control. The range is much wider compared with other five joints. In such a case, it is difficult to construct a sufficiently learned RNN with six units in the output layer. When the scales of the teaching signals for each unit in the output layer are quite different, the learning does not go well as a whole. That is the reason why six RNNs are independently designed. Six RNNs are learned for the trajectory following control, respectively.

At first, the down scaler vector $\boldsymbol{S} = \text{diag}\{S_1, \ldots, S_6\}$ is set to $\text{diag}\{1, \ldots, 1\}$. The learnings of six joints were respectively repeated 20000 times in order that the values of the error functions converged to almost zero. However, the learning of second joint could not be finished satisfactorily, i.e., the error did not decrease from the initial value. To overcome the problem concerning the learning, the second joint's teaching signals in the output layer were adaptively scaled down with

$$S_2 = \frac{\lambda}{\max(|\tau_2(k)|)} = \frac{5}{30} \fallingdotseq 0.17 \tag{15}$$

i.e., by giving $\boldsymbol{S} = \mathrm{diag}\{1, 0.17, 1, 1, 1, 1\}$, so that the error of the second joint could be reduced to almost zero as shown in Fig. ??(b). The errors of other joints are also shown in Fig. ??. Note that when the learned RNN for the second joint is used for a feedforward controller which can compensate the nonlinear terms such as gravity term $\hat{\boldsymbol{G}}(\boldsymbol{\theta})$ and Coriolis/centrifugal term $\hat{\boldsymbol{H}}(\boldsymbol{\theta}, \dot{\boldsymbol{\theta}})$ in Eq. (??), the output of the second joint must be scaled up with S_2^{-1} by multiplying $\boldsymbol{S}^{-1} = \mathrm{diag}\{1, 6, 1, 1, 1, 1\}$.

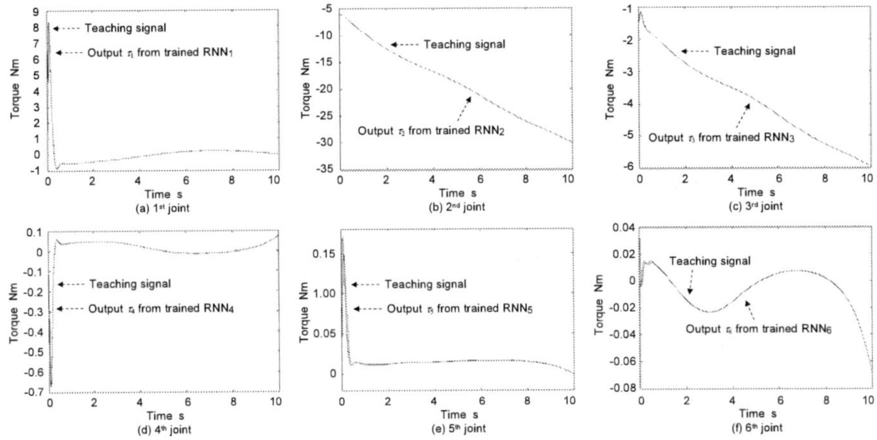

Figure 7. Learning results of joint driving torques for six joints.

2.3.3 Discussion

It is known that the form and position of sigmoid function can be changed by using a gain α and threshold β, respectively. The function and its differential are respectively represented by

$$f(X) = \frac{1}{1 + \exp\{-\alpha(X - \beta)\}} - 0.5 \tag{16}$$

$$f'(X) = \alpha f(X)\{1 - f(X)\} \tag{17}$$

If the gain α is set to 10, then the slope becomes steeper and the nonlinear range on horizontal axis becomes narrow as shown in Fig. **??**. Because the learning algorithm written by Eq. (**??**) includes the differential $f'(X)$, learning effectiveness can be expected only in the narrow range, i.e., activation area. Accordingly, when the absolute value of input to the sigmoid function is large, i.e., out of range, the updates of weights will not progress well. On the other hand, if α is set to 0.1, the slope becomes linear and gentle. The sigmoid function will lose the flexible characteristics of nonlinearity. Figure **??** illustrates the differentials with several α. The most important point is the adjustment of the input value, i.e., the error signal by using the down scaler $\boldsymbol{S}$. For example, if α is set to 1, it is desirable that the input values move within about ± 5 from the center as related in the painted area of Fig. **??**. In the learning process of the output layer of the second joint, the absolute value of the error between the teaching signal and output of the corresponding RNN_2 became much larger than about 5, so that the updates of weights and thresholds could not progress satisfactorily.

In this section, we have proposed a simple and adaptive learning method for the second joint, in which the teaching signals are scaled down to be mapped within about ± 5, i.e., into the activation area, on the horizontal axis shown in Fig. **??**. The effectiveness has been confirmed from the learning result of the second joint. Of course, a similar effectiveness will be performed by setting the gain α smaller and the learning coefficient η larger. However, the synergistic tuning according to the error signals in the output layer is not so easy, and also time-comsuming. By the way, it is common that antecedent membership functions are manually or automatically placed on support set according to actual fuzzy input signals when a fuzzy approach is incorporated in a control system. In some way, the proposed adaptive scaling down technique for error signals in output layer is similar to the design of such membership functions.

2.4 Advanced servo system using integrated RNNs

In this subsection, a servo control with the learned RNNs is introduced. The six RNNs independently learned are integrated for a feedforward controller as

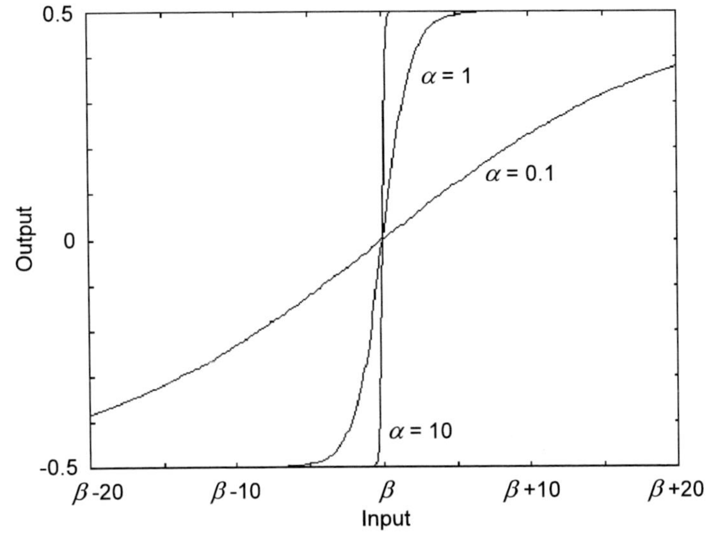

Figure 8. Sigmoid functions with different gains α and a threshold β.

shown in Fig. **??**. The RNNs act instead of the model-base portion, i.e., gravity term and Coriolis/centrifugal force term, in the computed torque control law given by Eq. (**??**). Figure **??** shows the block diagram of the control system including the integrated RNNs, in which the torque vector $\boldsymbol{\tau}_{NN}$ from the integrated RNNs is multiplied by $\boldsymbol{S}^{-1}$ to call back the original scales of teaching signals.

The control system is similarly applied to the trajectory following control problem as shown in Fig. **??**. Figures **??**(a)-(f) show the control results of the six joints, respectively. It is observed that each joint can be successfully controlled along the desired trajectory although there exist small delays between the desired joint angles and controlled ones. In this case, the elements of the output $\boldsymbol{\tau}_{FB}$ from the servo system without gravity term and Coriolis/centrifugal force term are shown in Fig. **??**(a)-(f), respectively. As can be seen from the small values of six joint driving torques, the integrated RNNs work well as a feedforward controller which can compensate both the gravity term and Coriolis/centrifugal force term.

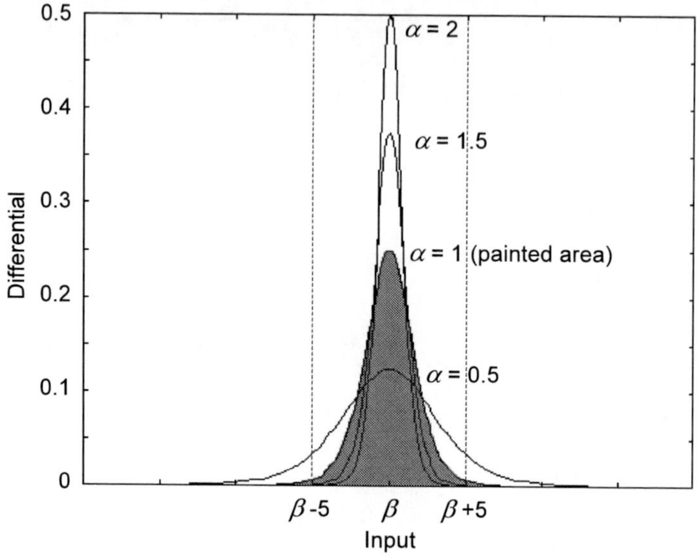

Figure 9. Differentials $f'(X)$ of sigmoid functions with different gains α.

3 Fine Gain Tuning for Model-Based Robotic Servo Controllers Using Genetic Algorithms

3.1 Background

Up to now, a large number of control strategies have been proposed and applied to various industrial fields. When we design a control system, one of the most complicated problems is that how the gains used in the control methods are tuned for each desired performance. In almost cases, the gains are experimentally tuned by trial and error based on the known model information. Recently, promising studies using genetic algorithms have been reported. For example, Deris, Omatu *et al.* considered the stabilization of an inverted pendulum which can be controlled by moving a cart in an intelligent way [?, ?]. To stabilize the pendulum, a PID controller which requires the determination of control gains is used. A neural network approach tunes the gains. Genetic algorithms are applied to find initial values of the connection weights of the neural network and initial values of PID gains. Also, the genetic algorithms-based optimized hybrid controller which is suitable for controlling both linear and nonlinear systems was proposed by Sun and Er [?, ?]. The controller consists of a linear PID

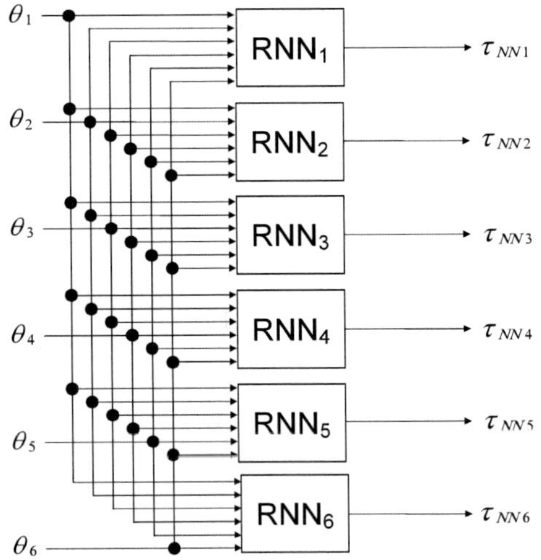

Figure 10. Integrated RNNs for a feedforward controller.

controller and a linear fuzzy logic controller, and employs a GA to facilitate optimal tuning of controller gains. Ge, Lee *et al.* proposed a class of energy-based position controllers for a kind of flexible spacecraft [?]. The feedback gains of the controller are tuned by a genetic algorithm optimization process to achieve good results for tip motion based on some suitable fitness functions. In [?], Tsuji, Hachino *et al.* described a design method of a controller by using genetic algorithms. The feedback gains are determined so that the response can coincide with the desired response. Besides, Senjyu, Arakaki *et al.* presented nonlinear excitation control for improving electric power system transient stability using a neural network (NN). The gains of the nonlinear excitation controllers modeled by NN are optimized by using genetic algorithms [?]. Before these researches, indeed, Davidor had already introduced genetic algorithms and its application to robotic trajectory optimization [?]. However, in these previous works using genetic algorithms, the search space narrowing through manual tuning process has not been efficiently used together.

Resolved acceleration control method [?] and computed torque control method are used for nonlinear control of industrial manipulators, which are composed of a model base portion and a servo portion. The servo portion is a closed loop with respect to the position and velocity. On the other hand, the model base portion has the inertia term, gravity term and centrifugal/Coriolis

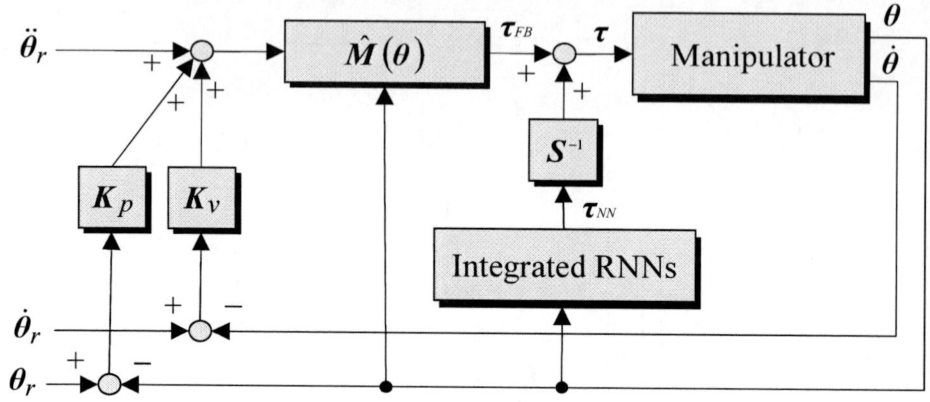

Figure 11. Block diagram of servo system comprising the integrated RNNs illustrated in Fig. ??.

term, which works for canceling the nonlinearity of manipulator. In order to realize high control stability, the position and velocity gains used in the servo portion should be tuned suitably.

In the latter half of the chapter, a simple but effective fine tuning method after the manual tuning process is introduced for the position and velocity feedback gains in the servo portion [?, ?, ?]. At the first step, search space for the gains is roughly narrowed down by a controller designer, e.g. while considering the critically damped condition. At the second step, the gains are finely tuned around the critically damped curve by using genetic algorithms. Genetic algorithms search for a better combination of the position and velocity gains. Simulations are conducted using a dynamic model of PUMA560 manipulator to validate the effectiveness of the proposed method.

3.2 Robotic Servo Controller

3.2.1 Resolved Acceleration Control

The dynamic model of a manipulator without friction term is generally given by

$$\boldsymbol{M}(\boldsymbol{\theta})\ddot{\boldsymbol{\theta}} + \boldsymbol{H}(\boldsymbol{\theta}, \dot{\boldsymbol{\theta}}) + \boldsymbol{G}(\boldsymbol{\theta}) = \boldsymbol{\tau} \tag{18}$$

where, $\boldsymbol{M}(\boldsymbol{\theta}) \in \Re^{6\times6}$ is the inertia term in joint space. $\boldsymbol{H}(\boldsymbol{\theta}, \dot{\boldsymbol{\theta}}) \in \Re^{6\times1}$ and $\boldsymbol{G}(\boldsymbol{\theta}) \in \Re^{6\times1}$ are the Coriolis/centrifugal term and gravity term in joint space,

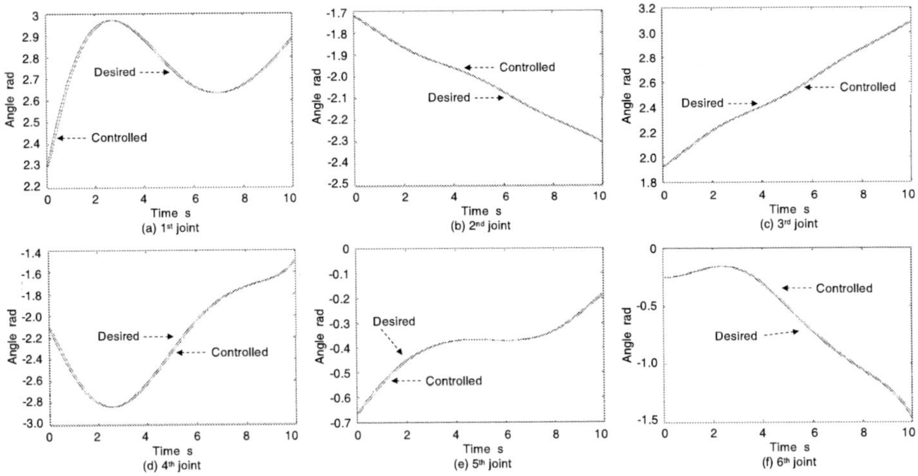

Figure 12. Control results of joint angles $\boldsymbol{\theta}$.

respectively. $\boldsymbol{\theta} \in \Re^{6\times 1}$, $\dot{\boldsymbol{\theta}} \in \Re^{6\times 1}$ and $\ddot{\boldsymbol{\theta}} \in \Re^{6\times 1}$ are the position, velocity and acceleration vector in joint coordinate system, respectively. $\boldsymbol{\tau} \in \Re^{6\times 1}$ is the joint driving torque vector.

In the case that the resolved acceleration control law is employed in the servo system of a manipulator, the desired position, velocity and acceleration vectors in Cartesian coordinate system are respectively given to the references of the servo system, so that the joint driving torque is calculated from

$$\begin{aligned} \boldsymbol{\tau} &= \hat{\boldsymbol{M}}(\boldsymbol{\theta})\boldsymbol{J}^{-1}(\boldsymbol{\theta}) \times \left[\ddot{\boldsymbol{x}}_r + \boldsymbol{K}_v\{\dot{\boldsymbol{x}}_r - \dot{\boldsymbol{x}}\} + \boldsymbol{K}_p\{\boldsymbol{x}_r - \boldsymbol{x}\} - \dot{\boldsymbol{J}}(\boldsymbol{\theta})\dot{\boldsymbol{\theta}}\right] \\ &+ \hat{\boldsymbol{H}}(\boldsymbol{\theta}, \dot{\boldsymbol{\theta}}) + \hat{\boldsymbol{G}}(\boldsymbol{\theta}) \end{aligned} \quad (19)$$

where, ˆ denotes the modeled term. $\boldsymbol{x} \in \Re^{6\times 1}$, $\dot{\boldsymbol{x}} \in \Re^{6\times 1}$ and $\ddot{\boldsymbol{x}} \in \Re^{6\times 1}$ are the position/orientation, their velocity and acceleration vectors in Cartesian coordinate system, respectively. $\boldsymbol{x}_r$, $\dot{\boldsymbol{x}}_r$ and $\ddot{\boldsymbol{x}}_r$ are the desired position/orientation, velocity and acceleration vectors, respectively. $\boldsymbol{K}_v = \mathrm{diag}(K_{v1}, \ldots, K_{v6})$ and $\boldsymbol{K}_p = \mathrm{diag}(K_{p1}, \ldots, K_{p6})$ are the feedback gains of velocity and position, respectively. $\boldsymbol{J}(\boldsymbol{\theta})$ is the Jacobian which gives the relation $\dot{\boldsymbol{x}} = \boldsymbol{J}(\boldsymbol{\theta})\dot{\boldsymbol{\theta}}$. Note that $\boldsymbol{\theta}$, $\dot{\boldsymbol{\theta}}$, $\boldsymbol{x}$ and $\dot{\boldsymbol{x}}$ in Eq. (**??**) are actual values, i.e., controlled variables. The nonlinear compensation terms $\hat{\boldsymbol{H}}(\boldsymbol{\theta}, \dot{\boldsymbol{\theta}})$ and $\hat{\boldsymbol{G}}(\boldsymbol{\theta})$ are calculated to cancel the nonlinearity and are effective to achieve a stable trajectory control. Figure **??** shows the block diagram of the resolved acceleration control method, in which

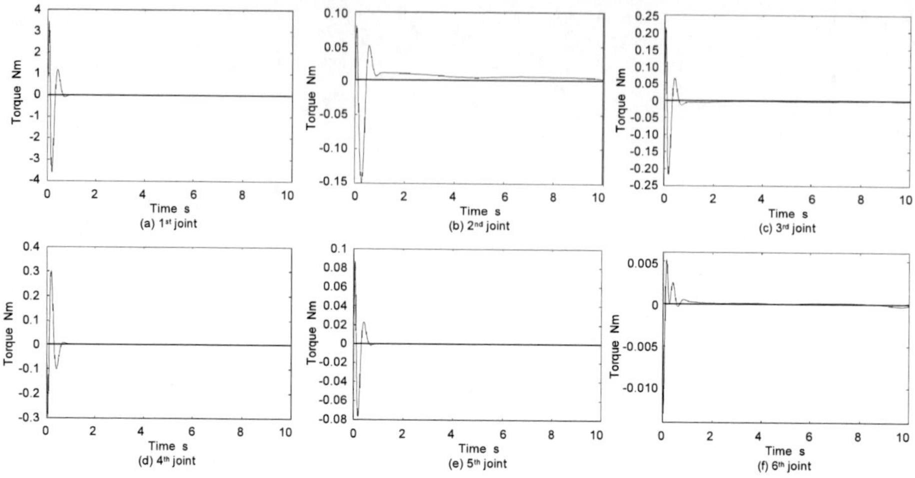

Figure 13. Joint driving torques $\boldsymbol{\tau}_{FB}$ generated from the servo system without gravity term and Coriolis/centrifugal force term as shown in Fig. ??.

$Fkine(\boldsymbol{\theta})$ is the function to obtain the forward kinematics.

3.2.2 Basic Gain Tuning Considering Critically Damped Condition

A basic gain tuning method considering the critically damped condition is explained giving an example. If it is assumed that the modeled terms ^ are exact, then the nonlinearity of the robot dynamics can be canceled. Substituting Eq. (??) or Eq. (??) into Eq. (??) gives the following second order error system.

$$\ddot{\boldsymbol{e}} + \boldsymbol{K}_v \dot{\boldsymbol{e}} + \boldsymbol{K}_p \boldsymbol{e} = 0 \tag{20}$$

where, $\boldsymbol{e} = \boldsymbol{x}_r - \boldsymbol{x}$ or $\boldsymbol{e} = \boldsymbol{\theta}_r - \boldsymbol{\theta}$. Accordingly, $\boldsymbol{K}_v$ and $\boldsymbol{K}_p$ are selected suitably, e.g., considering the critically dumped condition given by

$$K_{vi} = 2\sqrt{K_{pi}}\ (i = 1, \ldots, 6) \tag{21}$$

so that desirable responses can be obtained if under the ideal condition as given by Eq. (??). It should be noted, however, that there exists the nonlinearity such as Coulomb friction and viscous friction. The robotic dynamic model with the friction term is written by Eq. (??). Because $\hat{\boldsymbol{M}}(\boldsymbol{\theta})$, $\hat{\boldsymbol{H}}(\boldsymbol{\theta}, \dot{\boldsymbol{\theta}})$, $\hat{\boldsymbol{G}}(\boldsymbol{\theta})$ and $\boldsymbol{F}_r(\boldsymbol{\theta}, \dot{\boldsymbol{\theta}})$ bring out undesirable modeled treatment error, the velocity and position gains in the servo portion should be more finely tuned after the basic gain

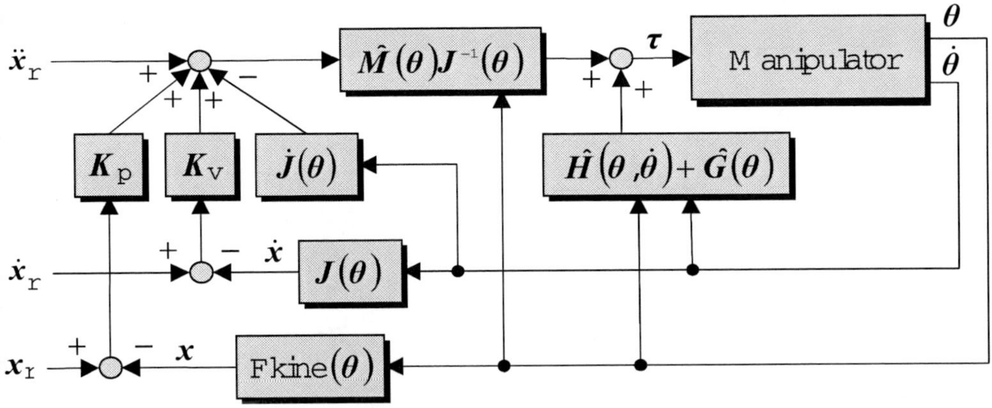

Figure 14. Block diagram of the resolved acceleration control method, where x_r, $\dot{x}_r$ and $\ddot{x}_r$ are the desired position/orientation, velocity and acceleration vectors in Cartesian coordinate system.

tuning. In the subsection 3.3, genetic algorithms are applied to search for more suitable gains.

3.2.3 Desired Trajectory

Simulations are carried out by using the dynamic model of the PUMA560 manipulator as shown in Fig. ??. For instance, in order to apply the resolved acceleration control method to the manipulator, the desired trajectory composed of x_r, $\dot{x}_r$ and $\ddot{x}_r$ in Cartesian coordinate system must be prepared. In like manner, the computed torque control method needs the desired trajectory composed of θ_r, $\dot{\theta}_r$ and $\ddot{\theta}_r$ in joint coordinate system. First of all, the desired trajectory in Cartesian coordinate system is designed as shown in Fig. ??, in which the manipulator moves from the initial pose to the final pose. In this case, the robot stretches the arm tip 500 mm. The manipulator is still in both initial and final poses. The desired trajectory x_r, $\dot{x}_r$ and $\ddot{x}_r$ are calculated through 4-1-4 order polynomial equation. θ_r, $\dot{\theta}_r$ and $\ddot{\theta}_r$ are transformed from x_r, $\dot{x}_r$ and $\ddot{x}_r$ by using the inverse kinematics, Jacobian $J(\theta)$ and sampling width Δt. We can design various trajectories with the combination of the acceleration time, constant velocity time and deceleration time.

An example that consists of desired joint angles, angle velocities and accelerations is shown in Fig. ??, which is needed for the computed torque control

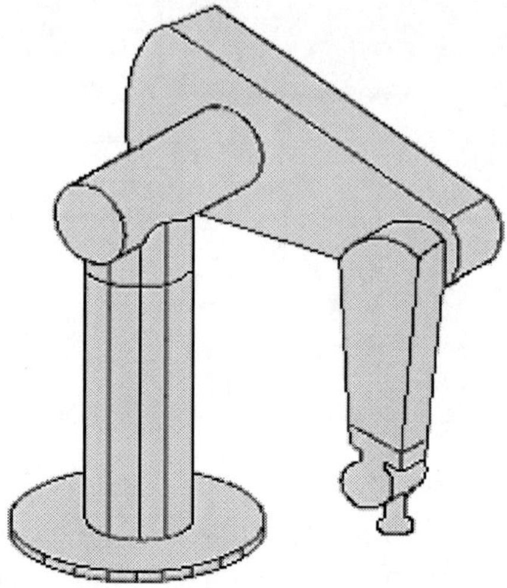

Figure 15. PUMA560 manipulator.

method to perform the motion shown in Fig. ??. To obtain satisfactory and safe control performance without falling a singularity, $\boldsymbol{K}_v$ and $\boldsymbol{K}_p$ are roughly tuned in advance with trial and error, considering the combination around the critically damped condition. We call this the initial manual tuning process. Two search spaces for $\boldsymbol{K}_v$ and $\boldsymbol{K}_p$ are obtained after the manual tuning process, so that $\boldsymbol{K}_v$ and $\boldsymbol{K}_p$ must be further tuned finely within the searched spaces to achieve a desirable motion without large overshoots and oscillations. In this subsection, we propose a systematic tuning method using genetic algorithms which can be applied to after the manual tuning process.

3.3 Fine gain tuning by using genetic algorithms

Gain tuning after designing a robotic servo controller is an important and indispensable task to realize a smooth motion control system. In this subsection, the gain tuning process consists of two steps. The first step is conducted to narrow down the search space for gains by operator's manual gain tuning while considering the circumference of the critically damped conditions given by Eq. (??). The ranges for gain search are narrowed down between the minimum values and the maximum values. The objective of the second step is to finely search for a

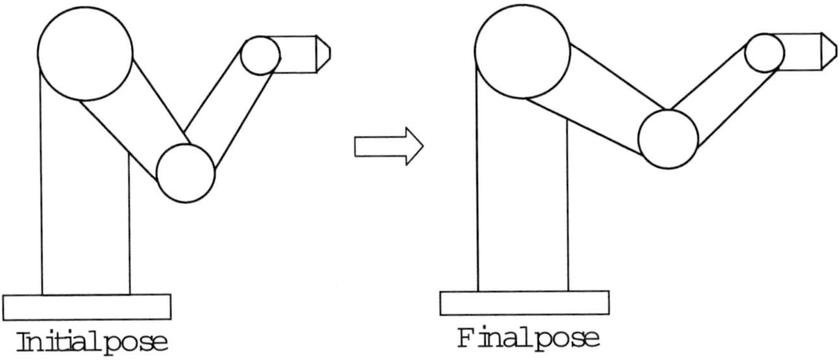

Figure 16. Trajectory following control problem.

more superior combination of K_{vi} and K_{pi} $(i = 1, \ldots, 6)$ within the search space by using genetic algorithms.

The genetic algorithm is used to optimize K_{vi} and K_{pi} which are the diagonal elements of $\boldsymbol{K}_v$ and $\boldsymbol{K}_p$. In order to apply genetic algorithms, K_{vi} and K_{pi} are transformed into a genotype. Diagonal elements K_{vi} and K_{pi} are encoded into 8 bits binary code and an individual is composed of 96 $(= 8 \times 12)$ bits. A population has 60 individuals. Each individual in initial population is generated within expectable range considered in the initial manual tuning process. For example, we have preliminarily confirmed by randomly conducted manual tuning that if each element of $\boldsymbol{K}_v$ and $\boldsymbol{K}_p$ in Eq. (**??**) is set to $50 \leq K_{vi} \leq 200$ and $10 \leq K_{pi} \leq 50000$ $(i = 1, \ldots, 6)$ then undesirable singularities don't tend to occur. The singularity is a typical result after the system becomes unstable.

The elite survivable strategy is adopted in the selection process, where superior six individuals can survive to next generation as elites. Crossover and mutation which are the representative GA operations are not applied to the genotypes of the elites. Other remained 54 individuals are yielded by a tournament selection. The genotype including the $\boldsymbol{K}_v$ and $\boldsymbol{K}_p$ is shown in Fig. **??**. Each individual is evaluated by applying Eq. (**??**) or Eq. (**??**) to the trajectory following control problem shown in Fig. **??**. The evaluation value E_v is calculated based on the square root of the sum of each directional error squared every sampling time, which is obtained by

$$E_v = \sum_{k=1}^{\frac{T}{\Delta t}} \sqrt{\sum_{i=1}^{3} \{e_i(k)\}^2} \tag{22}$$

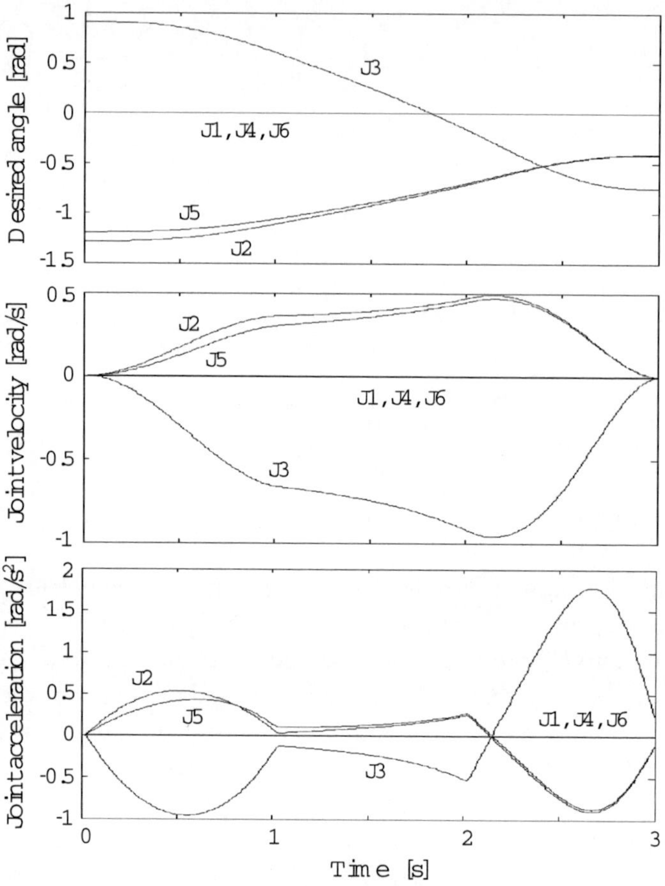

Figure 17. Desired joint angle $\boldsymbol{\theta}_r$, velocity $\dot{\boldsymbol{\theta}}_r$ and acceleration $\ddot{\boldsymbol{\theta}}_r$ in joint space.

K_{v1}	K_{v2}	K_{v3}	K_{v4}	K_{v5}	K_{v6}	K_{p1}	K_{p2}	K_{p3}	K_{p4}	K_{p5}	K_{p6}

Figure 18. Genotype of $\boldsymbol{K}_v$ and $\boldsymbol{K}_p$ which means an individual.

where the error $e_i(k)$ $(i = 1, 2, 3)$ is the x-, y- and z-directional position errors in Cartesian space. T and Δt are the simulation time and sampling period, respectively. This is a minimization problem, where individuals with smaller fitness E_v can survive as elites to the next generation.

3.4 Tuning Results

In this section, two types of robotic dynamics with and without the friction torque term are used for simulations. Simulations are carried out by using the dynamic model of PUMA560 manipulator on MATLAB system [?, ?].

Table 1. Parameters of GA operation in case of resolved acceleration control law

Population size	60
Number of elites	6
Selection	Tournament selection
Crossover	Uniform crossover (rate=87.5%)
Mutation	Random mutation (rate=4.17%)
Search space of K_{vi}	$0.001 \leq K_{vi} \leq 1000$
Search space of K_{pi}	$0.001 \leq K_{pi} \leq 50000$
Maximum generation	100

3.4.1 Robotic Dynamics without Friction Torque Term

First of all, the robotic dynamics without the friction torque term given by Eq. (??) is used.

(1) In case of resolved acceleration control law

Before applying the GA operation, we broadly investigated the characteristics of the trajectory following control, while changing the combination of $\boldsymbol{K}_v$ and $\boldsymbol{K}_p$. As the result, the ranges in which the robot could move without falling a singularity were respectively $0.001 \leq K_{vi} \leq 1000$ and $0.001 \leq K_{pi} \leq 50000$ $(i = 1, \ldots, 6)$, so that the search space for GA was limited within these ranges. The uniform crossover was employed to the individuals with a high probability rate 87.5% $(7/8)$ and the reverse action with mutation rate 4.17% $(1/24)$ is given to all bits forming 54 individuals. Other parameters for GA operations are tabulated in Table ??. As an example of the tuning effectiveness, the variation of minimum fitness is shown in Fig. ??. It is observed that more excellent individuals have been selected through the alternation of generations with GA operations. After 100 generations, the phenotype of the best individual is represented by

$$\begin{aligned} \boldsymbol{K}_v &= \mathrm{diag}(149, 183, 180, 185, 188, 82) \\ \boldsymbol{K}_p &= \mathrm{diag}(24765, 9510, 33275, 2902, 13725, 5882) \end{aligned}$$

Each of which is plotted with "∘" in Fig. ??. When the best individual was applied to the trajectory following control problem, E_v became the minimum value 0.5135.

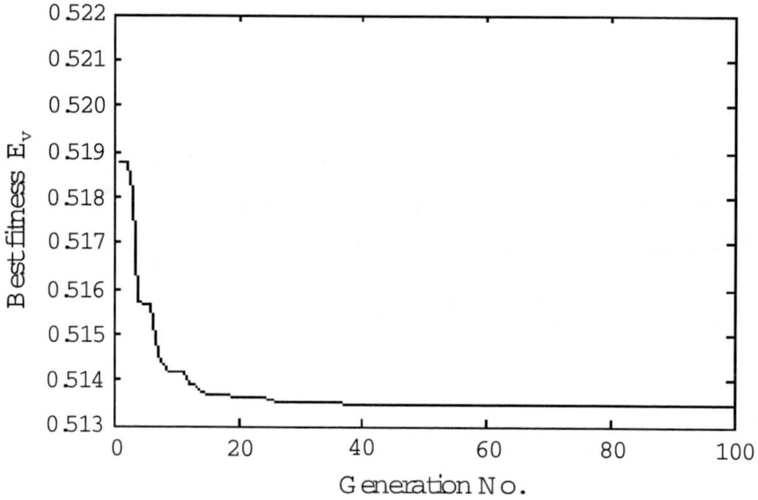

Figure 19. Evolutionary history in case of the resolved acceleration control law.

(2) In case of computed torque control law

Similar to the above case, the search space for GA was narrowed down to the ranges $70 \le K_{vi} \le 200$ and $100 \le K_{pi} \le 50000$. Also, the uniform crossover is applied to the individuals with a high probability rate 87.5% and the reverse action with mutation rate 6.67% (1/15) is given to the all bits forming 54 individuals except for 6 elites. Other parameters for GA operations are tabulated in Table ??. Figure ?? shows the result of evolutionary history, in which the computed torque control method is applied in robotic servo system.

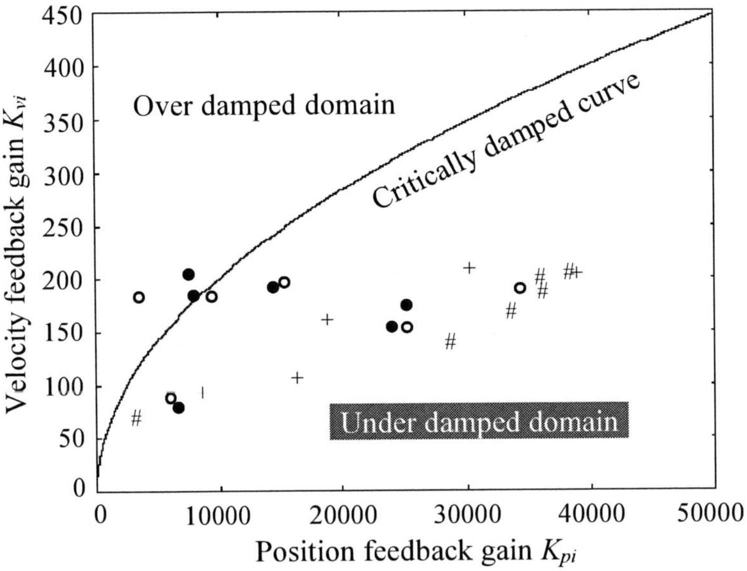

Figure 20. Relation between finely tuned position and velocity feedback gains obtained through GA operations.

After 100 generations, the phenotype of the best individual is represented by

$$\begin{aligned} \boldsymbol{K}_v &= \text{diag}(193, 196, 168, 70, 185, 138) \\ \boldsymbol{K}_p &= \text{diag}(35980, 38490, 33470, 3129, 35721, 28969) \end{aligned}$$

Each element of which is plotted with "♯" in Fig. ??. When the best individual was applied to the trajectory following control, E_v became the minimum value 0.5109.

3.4.2 Robotic Dynamics with Friction Torque Term

Next, the robotic dynamics with the friction torque term given by Eq. (??) is applied.

(1) In case of the resolved acceleration control law

Figure ?? shows the result of evolutionary history, in which the resolved acceleration control law was employed in the robotic servo system. The robotic dynamics used in simulation is written by Eq. (??). The search space for

Table 2. Parameters of GA operation in case of computed torque control law

Population size	60
Number of elites	6
Selection	Tournament selection
Crossover	Uniform crossover (rate=87.5%)
Mutation	Random mutation (rate=6.67%)
Search space of K_{vi}	$70 \leq K_{vi} \leq 200$
Search space of K_{pi}	$100 \leq K_{pi} \leq 50000$
Maximum generation	100

GA operation was narrowed down to the ranges $0.001 \leq K_{vi} \leq 220$ and $0.001 \leq K_{pi} \leq 40000$, again. Also, crossover rate and mutation rate were set to 87.5% and 4.17%, respectively. After 100 generations, the phenotype of the best individual is represented by

$$\boldsymbol{K}_v = \mathrm{diag}(157, 204, 171, 181, 192, 76)$$
$$\boldsymbol{K}_p = \mathrm{diag}(23137, 7412, 24471, 7922, 13725, 5882)$$

The relations are plotted with "•" in Fig. ??. When above gains were given, the minimum fitness E_v became 0.5241 by Eq. (??).

(2) In case of the computed torque control law

Figure ?? shows the result of evolutionary history, in which the computed torque control method was applied in robotic servo system. The search space for GA was narrowed down to the ranges $70 \leq K_{vi} \leq 220$ and $100 \leq K_{pi} \leq 50000$, again. Also, the crossover rate and mutation rate were set to 87.5% and 6.67% (1/15), respectively. After 100 generations, the phenotype of the best individual is represented by

$$\boldsymbol{K}_v = \mathrm{diag}(198, 196, 199, 102, 91, 156)$$
$$\boldsymbol{K}_p = \mathrm{diag}(30561, 39365, 39365, 16123, 8904, 19116)$$

The positions of the elements are plotted with "+" in Fig. ??. When above gains were given, the minimum fitness E_v became 0.5097 by Eq. (??). Figure ?? shows the simulation results that are performed by the gains with the minimum fitness. The upper figure in Fig. ?? shows the controlled angle trajectories of joints 2 and 3. The second figure similarly shows the angle velocities. The third

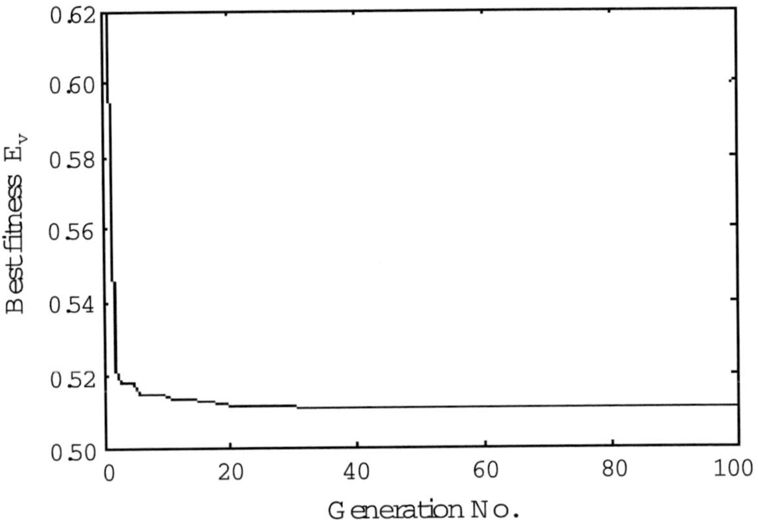

Figure 21. Evolutionary history in case of the computed torque control law.

and lower figures respectively illustrate the friction torques acting at the joints 2 and 3 which are given by Eq. (**??**). For example, it can be observed from the second figure in Fig. **??** that $\text{sign}(\dot{\theta}_2)$ and $\text{sign}(\dot{\theta}_3)$ are frequently changed around the start time and end time. According to the changes of $\text{sign}(\dot{\theta}_2)$ and $\text{sign}(\dot{\theta}_3)$, the large influence of Coulomb friction can be also observed around the corresponding time area in the third and lower figures.

Figure **??** shows the relation of the critically damped curve based on Eq. (**??**) and elite gains acquired by the GA operation. Summarizing the simulation results, it is observed that the best combination of K_{vi} and K_{pi} among obtained through the GA operation almost has the under damped condition given by

$$K_{vi} < 2\sqrt{K_{pi}} \; (i = 1, \ldots, 6) \tag{23}$$

This suggests that under damped situation is suitable for such a robotic trajectory following control with time-varying references as shown in Fig. **??**.

4 Conclusions

In this chapter, two types of intelligent control systems have been introduced. One is the adaptive learning technique for large-scale teaching signals. Inte-

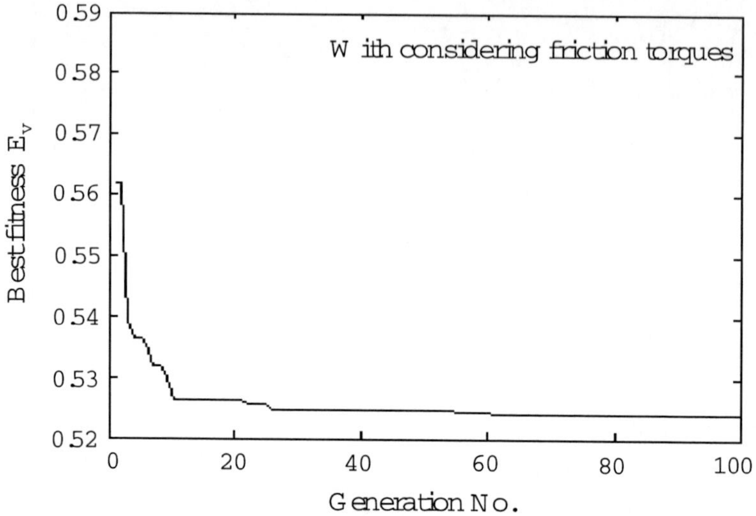

Figure 22. Evolutionary history in case that the resolved acceleration control law was employed.

grated recurrent neural networks (RNNs) have been proposed for PUMA560 manipulator to feedforwardly control a trajectory. A basic sigmoid function is used in each unit. The feedforward controller using the integrated RNNs approximately includes gravity compensator and Coriolis/centrifugal force compensator. In order to adaptively learn such large-scale teaching signals processed in back propagation algorithm as have absolute values over activation area of sigmoid function, the proposed RNNs scale down the teaching signals into the activation area. When the integrated RNNs are used, the output from the RNNs have only to be scaled up inversely. The adaptive learning technique by using the scale down factor and the control effectiveness of integrated RNNs have been evaluated through simulations using the dynamic model of PUMA560 manipulator. Consequently, a promising result has been confirmed compared with the conventional intuitive learning technique in which the gain of sigmoid function and learning coefficient are synergistically selected by trial and error. As for future work, it is being planned to apply the proposed adaptive learning technique to the feedback error learning [?, ?] which is expected to be effective so as to realize online learning of the integrated RNNs.

The other is the fine gain tuning technique for model-based robotic servo

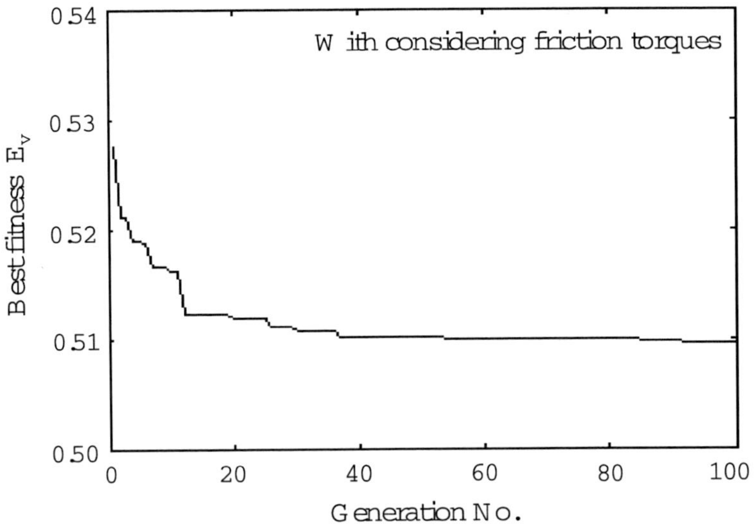

Figure 23. Evolutionary history in case that the computed torque control law was employed.

controllers using genetic algorithms. A fine gain tuning method using genetic algorithms, applied after the manual tuning process, has been proposed for the resolved acceleration control method and computed torque control method. These two control methods are respectively known as one of the nonlinear control strategies for industrial manipulators and are composed of a model base portion and a servo portion. Each servo portion is a closed-loop system with respect to the position and velocity. The results described in the latter half of the chapter are summarized as follows. First of all, the search spaces for the gains are narrowed down by the manual tuning considering the neighborhood of critically damped condition. After the manual tuning process, genetic algorithms search for a better combination of the position and velocity feedback gains within the roughly narrowed search spaces. Simulation results have demonstrated that finely tuned gains with minimum fitness allow the PUMA560 manipulator to perform stable trajectory following control under unknown nonlinearity such as joint friction torques.

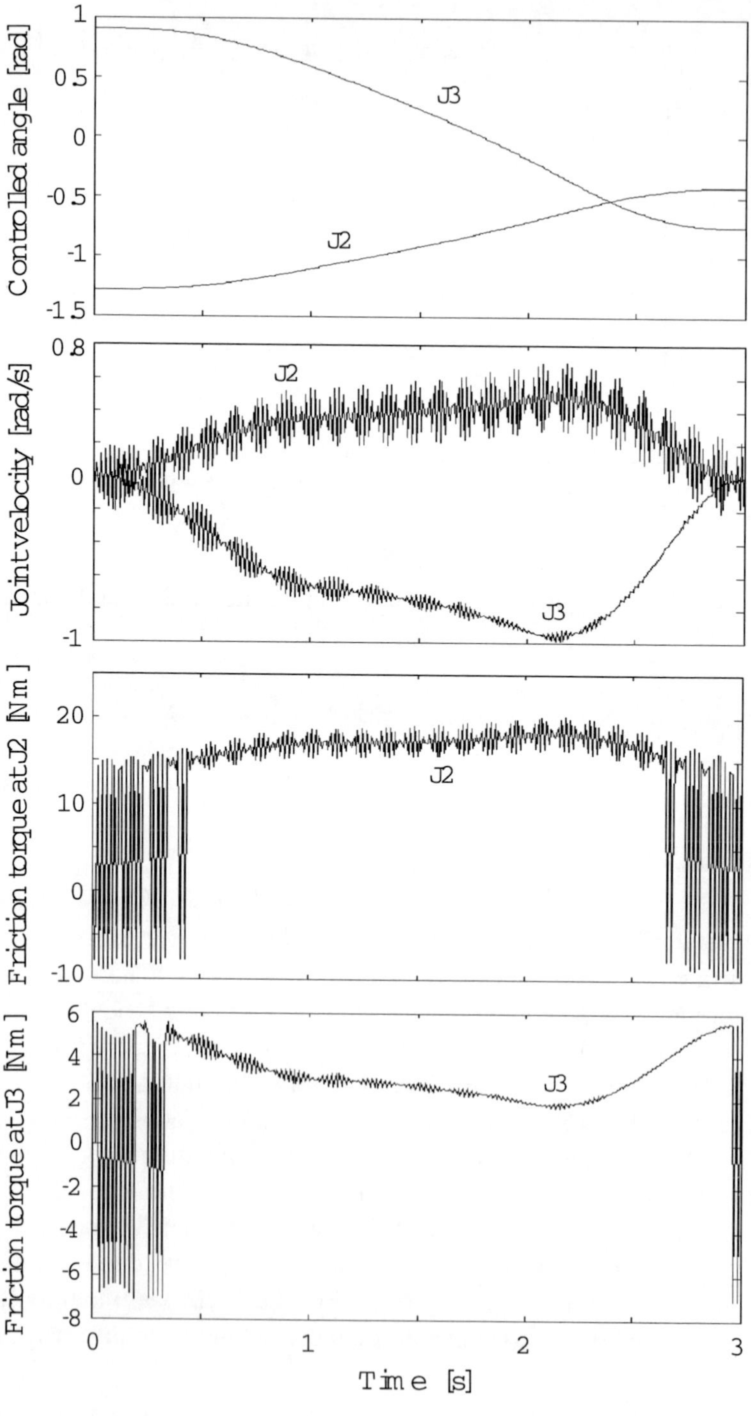

Figure 24. Simulation result in case of using the gains with minimum fitness.

References

[1] Nagata F., Watanabe K., Hase T., Haga Z., Omoto M., Tsuda K., Tsukamoto O., Komino M., Kusumoto Y. Application of open architectural industrial robots and its simulation technique. *International Journal of Computer Research*, 2008, vol.17, nos. 1–2, pp. 1–40.

[2] Chen C., Trivedi M. M., Bidlack, C. R. Simulation and animation of sensor-driven robots. *IEEE Trans. on Robotics and Automation*, 1994, vol. 10, no. 5, pp. 684–704.

[3] Benimeli F., Mata V., Valero F. A comparison between direct and indirect dynamic parameter identification methods in industrial robots. *Robotica*, 2006, vol. 24, no. 5, pp. 579–590.

[4] Nagata F. Simulation technique of velocity-based discrete-time control system with intelligent control concepts for open architectural industrial robots. *The Open Automation and Control Systems Journal*, 2008, vol. 1, pp. 31–43.

[5] Han J., Moraga C. The Influence of the sigmoid function parameters on the speed of backpropagation learning. *Lecture Notes In Computer Science*, 1995, vol. 30, pp. 195–201.

[6] Torras C. Robot adaptivity. *Robotics and Autonomous Systems*, 1995, vol. 15, nos. 1–2, pp. 11–23.

[7] Hasan A. T., Hamouda A. M. S., Ismail N., Al-Assadi H. M. A. A. An adaptive-learning algorithm to solve the inverse kinematics problem of a 6 D.O.F serial robot manipulator. *Advances in Engineering Software*, 2006, vol. 37, no. 7, pp. 432–438.

[8] Horng J. H. Hybrid MATLAB and LabVIEW with neural network to implement a SCADA system of AC servo motor. *Advances in Engineering Software*, 2008, vol. 39, no. 3, pp. 149–155.

[9] Huang S. N., Tana K. K., Lee T. H. Neural network learning algorithm for a class of interconnected nonlinear systems. *Neurocomputing*, 2009, vol. 72, nos. 4–6, pp. 1071–1077.

[10] Nagata F., Watanabe K. Adaptive learning with large variability of teaching signals for neural networks and its application to motion control of

an industrial robot. *International Journal of Automation and Computing*, 2011, vol. 8, no. 1, pp. 54–61.

[11] Paul R. P. *Modeling, trajectory calculation and servoing of a computer controlled arm*. Technical Report AIM-177, Stanford University Artificial Intelligence Laboratory, 1972.

[12] Craig J. J. *Introduction to ROBOTICS —Mechanics and Control Second Edition—*. Reading MA: Addison Wesley Publishing Company, 1989.

[13] Corke P. A robotics toolbox for MATLAB. *IEEE Robotics & Automation Magazine*, 1996, vol. 3, no. 1, pp. 24–32.

[14] Corke P. MATLAB toolboxes: robotics and vision for students and teachers. *IEEE Robotics & Automation Magazine*, 2007, vol. 14, no. 4, pp. 16–17.

[15] Deris S., Omatu S., Kitagawa K. Stabilization of inverted pendulum by the genetic algorithm. *Procs. of 1995 IEEE International Conference on Systems, Man and Cybernetics*, pp. 4372–4377, 1995.

[16] Omatu S., Deris S. Stabilization of inverted pendulum by the genetic algorithm. *Procs. of 1996 IEEE International Conference on Evolutionary Computation*, pp. 700–705, 1996.

[17] Sun Y. L., Er M. J. Optimal hybrid control of linear and nonlinear systems. *Procs. of the 40th IEEE Conference on Decision and Control*, pp. 1669–1674, 2001.

[18] Er M. J., Sun Y. L. Hybrid fuzzy proportional-integral plus conventional derivative control of linear and nonlinear systems. *IEEE Transactions on Industrial Electronics*, 2001, vol. 48, no. 6, pp. 1109–1117.

[19] Ge S. S., Lee T. H., Hong F. Robust controller design with genetic algorithm for flexible spacecraft. *Procs. of the 2001 Congress on Evolutionary Computation*, pp. 1033–1039, 2001.

[20] Tsuji T., Hachino T., Oguro R., Umeda N., Takata H. A control design of robotics using the genetic algorithm. *Artificial Life and Robotics*, 1998, vol. 2, no. 1, pp. 24–27.

[21] Senjyu T., Arakaki T., Uezato I. Stabilization control for multi-machine power system by nonlinear state feedback control using neural network. *Procs. of IEEE Power Engineering Society 1999 Winter Meeting* , pp. 622–627, 1999.

[22] Davidor Y. Genetic Algorithms and Robotics –A Heuristic Strategy for Optimization–. World Scientific Publishing, 1991.

[23] Luh J. Y. S., Walker M. H., Paul R. P. C. Resolved acceleration control of mechanical manipulator. *IEEE Trans. Automatic Control*, 1980, vol. 25, no. 3, pp. 468–474.

[24] Nagata F., Kuribayashi K., Kiguchi K., Watanabe K. Simulation of fine gain tuning using genetic algorithms for model-based robotic servo controllers. *Procs. of the 2007 IEEE International Symposium on Computational Intelligence in Robotics and Automation* , pp. 196–201, 2007.

[25] Nagata F., Okabayashi I., Matsuno M., Utsumi T., Kuribayashi K., Watanabe K. Fine gain tuning for model-based robotic servo controllers using genetic algorithms," *Procs. of the 13th International Conference on Advanced Robotics*, pp. 987–992, 2007.

[26] Nagata F., Watanabe K. Fine Gain Tuning for model-based robotic servo controllers using genetic algorithms. *Advances in Computer Science and Engineering*, 2008, vol. 2, no. 3, pp. 201–219.

[27] Kawato M. The feedback-error-learning neural network for supervised motor learning. *Advanced Neural Computers*. pp. 365–373; North-Holland: Elsevier, 1990.

[28] Nakanishi J., Schaal S. Feedback error learning and nonlinear adaptive control. *Neural Networks*, 2004, vol. 17, no. 10, pp. 1453–1465.

In: Industrial Control Systems
Editors: R. Gilbert and A. Schultz
ISBN: 978-1-61209-988-0

Chapter 6

MODELING AND CONTROL OF INDUSTRIAL SYSTEMS USING GLOBAL AUTOMATA

Vasileios Deligiannis*[*] *and Stamatis Manesis
University of Patras, Rion – Patras, Greece

ABSTRACT

In this work, Global Automata are presented, as a method offering the convenience of modeling various types of industrial systems without any restrictions on systems properties. Their structure has common characteristics with several types of automata, such as the control graph with a finite set of states and transitions between those states. Global automata can be used for modeling hybrid systems handling both discrete and real valued variables combining flow, invariant and guard conditions from hybrid automata, with clock constraints and delayed inputs from timed and PLC automata. In addition, new modeling parameters, as reset table at each transition and hierarchical classification of executable events at each state, are introduced. Application independence derives from the fact that they are a super-set of every other type of automata and hence are less application depended compared to any of them.

Based on Global Automata, some tools were developed in order to increase their modeling power. First tool is the ability of state

---

[*] First-Corresponding Author: Vasileios Deligiannis, Dr. Electrical & Computer Engineer, Systems & Measurement Laboratory, Electrical & Computer Engineering Dept., University of Patras, 26500, Rion – Patras, Greece, Tel. +30 6977 239855, e-mail: vdeligiannis@gmail.com

aggregation, which generally means the merging of two or more states in order to produce a new "super-state". State aggregation is a powerful tool to avoid state space expansion. The second developed tool is automata composition and is mainly used in cases where ad-hoc modeling is not a practical solution, since it allows the construction of large system model using simple models of all system's components. Finally, global automata allow the development of hierarchical models, consisting of two or more automata. In such case, the automaton of the higher level has the role of supervisor feeding the automata of lower level with data, reference values etc.

Two implementation tools for programming industrial controllers are also presented. The first tool is a synthesis tool for translating a global automaton into executable code of a programming language (e.g. C, Matlab). Tool usage is important for simulation and verification but is not limited on this. Contemporary industrial systems are equipped with Supervisory Control and Data Acquisition Systems (SCADA), which have embedded functions of running code. Consequently, an industrial system modeled as a global automaton can be controlled by the executable code running in a SCADA station. However in industry the dominant controller is the Programmable Logic Controller, which is a special aim computer suitably built for the application control tasks. The IEC 61131-3 Programming Norm describes all the well-known languages for programming PLCs and the second tool is a synthesis tool for implementing global automata in PLCs based on this programming norm.

Global automata, as a modeling method, and all the above mentioned tools are illustrated through representative examples. The presented examples are either theoretical or real systems implemented in the laboratory or even real production systems.

1. Introduction

The importance of modelling and simulation for the design of high quality industrial systems in a limited amount of time is generally acknowledged. But, which is the right definition of an industrial system? An industrial production line typically consists of various kinds of equipment (e.g. sensors, actuators, robots, valves, NC machines etc.) under the control of either centralized supervisor or decentralized controllers capable of performing a specific set of operational functions. From a planning and control perspective, an industrial production line can be seen as a dynamic system whose states evolve according to the occurrence of abrupt physical events, thus exhibiting the characteristics of a *discrete event system* (DES) [1]. In another point of view, systems of industrial relevance are governed by discrete state digital controllers, whose internal state transitions are triggered by the value of some measured continuous physical quantity (temperature, speed, fluid rate, etc.). Such systems, which are described by both discrete and continuous variables, are usually referred to as hybrid systems or more accurate as *hybrid control systems* [2, 3].

Studying industrial systems by simulation enables the designer to study their dynamic behaviour and determine characteristics of the system [4]. Especially today, where lifetime of industrial products decreases, while worldwide competition increases, the question of how correct models can be developed in a limited amount of time arises. Therefore, necessity of formal methods occurs [5, 6]. Formal methods allow a thorough analysis of the possible behaviours of a system and rigid proving of system properties in verification and validation. During the past decades, many methods have been proposed and examined from researchers for modelling industrial systems. Such methods are Petri nets and several types of automata. Petri nets [7], since their introduction in the early 1960's, have become a powerful tool for modelling and analysis of dynamical DESs. In contrast with scientists' expectations, Petri nets did not meet success when proposed for closed loop control, primarily because there was not a clear separation of plant and controller. On the contrary, automata [8, 9, 10, 11, 12] seem to be the right tool for solving this control problem. In general, an automaton is an intuitive and natural description of a DES, which reconstructs the system in a schematic way. Finite automata [11] are probably the simplest mathematical abstractions for modelling DESs. Finite automata model DESs with a finite number of states and transitions between those states. An extension of this classical notion, which models the coupled interaction of discrete events and continuous dynamical systems, are hybrid automata. Hybrid automata [10, 12, 13, 14, 15] have been proposed as a formal method for modelling hybrid systems and probably are the most well studied type of controlled automata. Another superset of finite automata are timed automata [9, 16], in which each transition is restricted by an additional set of clocks. Timed automata's subclass, PLC automata [8, 17], was first introduced in 1997 by Dierks as a useful tool for the description of distributed real time systems. PLC automata differ from timed automata, since they are restricted to only one clock and certain types of constraints on it. All these methods, despite the academic efforts, did not meet wide acceptance for industrial use, mainly because they are application depended, or more accurate domain depended. Except from the various automata types, some other modelling tools can be found in bibliography, such as CHARON, Ptolemy and Shift. These methods were, in general, introduced for design, modelling and simulation of hybrid systems but without specific industrial orientation.

In [18] *Global Automata* are introduced as a new method offering the convenience of modelling various types of industrial systems without any restrictions on systems properties. Their structure has common characteristics with several types of automata, such as the control graph with a finite set of states and transitions between those states. Global automata can be used for modelling hybrid systems handling both discrete and real valued variables combining flow, invariant and guard conditions from hybrid automata, with clock constraints and delayed inputs from timed and PLC automata. In addition, new modelling parameters as reset table at each transition and hierarchical classification of executable events at each state are introduced. Application independence derives from the fact that they are a super-set of every other type of automata and hence are less application depended compared to any of them.

Another bottleneck, relevant to industrial systems, is that real applications work with very large state spaces in comparison with academic papers. Typical academic projects present examples containing some tens of states and events but, as shown in [19], industrial projects contain *tens of thousands* of states. Hence, it is clear that building from scratch an industrial system's model is impracticable. Especially in contemporary industries, where interconnection of heterogeneous components in order to create a new homogeneous product, is needed. In such cases, composition of several subsystems is imperative since all the subsystems are different links of the same productive chain. Additionally, frequently one would have to design a new component that, combined with an existing part of a system, conforms to the given overall specifications [20]. Solution is again the same, composition of new part's model with existing system to produce whole system's model for simulation and verification of the desirable performance. Consequently, composition of two independent models is very useful on modelling industrial production systems. Global automata have the ability of composition, as also conventional automata types do [11, 21, 22, 23, 24]. But, composition results in state space sudden increase, since composing two automata, with m and n states respectively, produces a new automaton with $m \times n$ states [3, 25]. Global automata offer us a very powerful tool to avoid state space expansion taking advantage of their capabilities. One of them is the ability of state aggregation which generally means the merging of two or more states in order to produce a new super-state. Both, composition and state aggregation are presented analytically below in this paper.

The next two paragraphs present a brief literature review, the first one on the need of modeling and its various methods, while the second presents several tools/methods for automatic production of executable code for PLCs.

1.1. Modeling Industrial Systems – A Literature Review

Modeling and simulation are generally acknowledged of high importance for designing high quality industrial systems in a limited amount of time. Some thoughts on model equivalence are presented in [26], where an approach to understanding equivalence of behavior of discrete event simulation models is presented. It uses a technique from the concurrency world, Milner's Calculus of Communicating Systems, which seems to be advantageous in comparison to previous work. Concerning automata theory, many works have been published on modeling industrial systems and not only. Automata can be used for modeling in the field of embedded systems. Specifically, timed automata have been used to model and to analyze timeliness properties of embedded system architectures [27]. The presented case study is inspired by industrial practice and describes an in-car radio navigation system. Finite automata, automata's fundamental form, are used in [28] as the base of an object oriented design approach built around the formal theory of supervisory control. The application domain of the presented simulation framework is the flexible manufacturing systems. Time automata are also used as the modeling tool for modeling and analysis of fault latency. In [29] an effective way of modeling complex systems through identified functioning modes is presented. The whole approach is oriented for application to manufacturing systems and the simulation gives good results. Moreover, timed automata are used to model complex production scheduling problems such as the job-shop and the open-shop scheduling problems. In [30] the implementation of a method for production scheduling, using advance model checking tools, such as UPAAL, is presented. Another approach for solving job-shop scheduling problems based on reachability analysis for timed automata is given in [31]. The paper further improves the efficiency of the algorithm by excluding suboptimal or redundant solutions from the search space. Experimental results for several benchmark problems show that in a given computation time the proposed method produces better schedules. Supavatanakul et al [32] deal with the problem of fault diagnosis in discrete event systems which are represented by timed automata. The proposed diagnostic algorithm identifies faults in the system by comparing the input and output sequences with the timed automaton. In comparison with other methods, the given approach does not require any knowledge of the initial state of the system and hence it can be applied spontaneously to the timed automaton. The algorithm is applied for fault detection and isolation of the DAMADICS actuators and all the relative results are presented. Modeling and control of automated manufacturing systems is presented by Balduzzi and Kumar [15] based on hybrid automata. There, each system's basic service is reconstructed as an elementary hybrid automaton and after composing all these automata, one can have the model of the whole manufacturing system. Guo et al [33] presented a new prototype system oriented to manufacturing systems that is SimOpt. SimOpt consists of a virtual simulation modeling subsystem based on the object-oriented modeling technique and an optimization subsystem, which uses a new hybrid optimization algorithm. Authors' willing is to further improve the whole system so as to develop better performance for complex manufacturing systems. Another approach on modeling and analysis of complex systems is presented in [34]. The effectiveness of the proposed model is illustrated by applying it to the modeling and simulation of an industrial plant in France. In this work, a generic framework for naturally distributed and complex systems based on Holonic Multi-Agent paradigm is described. Finally, a detailed literature review on manufacturing systems can be found in [35] relatively to productivity measurement and improvement. It also includes a survey of available commercial tools to measure manufacturing systems performance and to perform factory level diagnostics.

1.2. Automatic PLC Code Generation – A Literature Review

As already clearly highlighted, the Programming Logic Controller is the governing controller in the industry and since, the development of PLC programs is an exhaustive and expensive activity [36, 37]; there are many approaches of automatic PLC code generation from system's model. In literature, such approaches can be found, which use various modelling methods and programming languages. In [38] two ways to solve the implementation problem are presented: a) top down approach, from model to auto generation of control logics, or b) bottom up approach, starts with manual logic programming and formalises to some software based techniques like converting the PLC code to automata and verifies using software based simulations and algorithm techniques.

Supremica has algorithms for verification and synthesis [39]. In addition, it can automatically generate code for a number of languages including ANSI C and IEC 61131 languages such as ST and IL. The paper [40] focuses on the implementation stages where the derived automata models are translated into program code which can be downloaded into a standard industrial PLC. This way, the development and implementation phases of industrial automation projects are significantly shortened. Another approach is given in [41], which presents a transformation translating the given automata into executable PLC code.

An approach to implement a Programmable Logic Controller from real-time software components following the International Industrial Standard IEC61499 is given in [42]. In [43], a method is presented for modeling, verification and automatic programming of PLC controllers. The method offers a formal model of requirements based on UML state machine, and a technique for generating program code. The UML state machine is used since it provides a widely accepted means for writing a specification at a suitable high level of abstraction. Such an abstract specification can be validated by the user, verified by means of a model-checker and translated automatically into a program code, in one of the standardized IEC 61131-3 languages.

An attempt to implement Supervisory Control Theory for discrete event systems in PLCs is introduced in [44]. The resulting PLC program is modular in structure, making it hand able for modification and error detection. Another paper which demonstrates how PLC-code for process observation can be automatically generated from a discrete event process model expressed as either PLC-code or as extended finite automata is [45]. Using the PLC languages of the IEC 61131-3 standard [46] for process modeling, narrows the gap between the pragmatic industry and the formal academia. Programmable Logic Controllers are also the target machine for implementing supervisors in [47]. In this work, a new conversion algorithm which directly transforms a supervisor represented by a finite automaton to a Ladder Logic Diagram is developed. Finally, in [48] an attempt to reduce the development time of PLC by automating the task of controller logic design and implementation is shown. As a modeling method, the t-MPSG, an extended finite state automaton, is used. It is developed for generating execution module for the real-time shop floor controller system. The t-MPSG supervises the flow of part from one state to another state considering the precondition of the system. It can be used to model the controller logic, which can be consequently translated to textual code, and finally translated into IEC standard PLC code.

Every presented method for automatic production of PLC code has the same main goal, which is to reduce the development time and the improvement of productivity in industrial manufacturing systems. In [35], a review of literature on manufacturing systems productivity measurement and improvement is shown. This review has been summarized under four categories; they are Operations Research-based methods, system analysis-based methods, continuous improvement methods and performance metrics-based methods. A survey of commercial tools available to measure manufacturing system performance is also performed.

2. Automata Theory

This section is dedicated to the definitions of the four main types of automata, which can be found in the bibliography [8, 9, 10, 11, 12, 13]. Here, the characteristics of each type are presented.

2.1. Finite Automata

Finite state automata are probably the simplest mathematical abstractions of discrete event systems with a finite number of states and transitions between those states. In every time moment, the automaton is always in one state. A transition between two states is caused by an event. Each event is considered instantaneous and hence, it cannot be interrupted. In addition, two events cannot occur simultaneously and only the present state's events are executable.

Definition 2.1. Finite Automata

A deterministic finite state automaton is described by a 4-tuple: $A = \langle Q_A, \Sigma_A, i_A, \delta_A \rangle$, where:

- Q_A is a finite nonempty set, called *set of states*,
- Σ_A is a finite nonempty set of events, called *alphabet,*
- $i_A \in Q_A$ is the initial state, which is indicated by an arrow in the automaton and

- $\delta_A: Q_A \times \Sigma_A \rightarrow Q_A$ is the *transition function*, which describes the transition between two sequential states, caused by an event.

Figure 2.1 depicts a simple switch automaton model, in which q_1 and q_2 are states and *a, b* are events, while q_1 is the initial state.

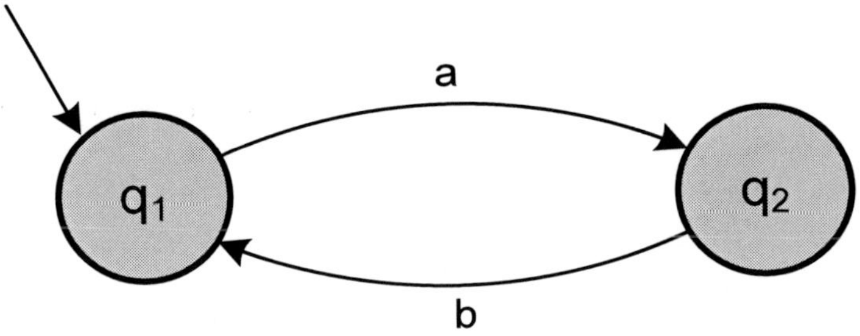

Figure 2.1. Finite automaton.

Some other researchers use a similar definition of finite state automata adding one more parameter, which is the *set of marked states* $M_A \subseteq Q_A$. These marked states are generally the desired states and are used to distinguish when a sub-task is done. The set of marked states are referred also, in bibliography, as *set of final states*.

Frequently, two automata are brought together in the intention to interact with each other. The automata's common events guide the interaction and must have the same properties. The interaction of two automata A and B is modeled by the *synchronous composition* $A \| B$, where:

$$A \| B = \langle Q_A \times Q_B, \Sigma_A \cup \Sigma_B, \langle i_A, i_B \rangle, \delta_{A\|B} \rangle,$$

$$\text{with } \delta_{A\|B}(\langle q_A, q_B \rangle, \sigma) = \begin{cases} \delta_A(q_A, \sigma) \times \delta_B(q_B, \sigma), \sigma \in \Sigma_A \cup \Sigma_B \\ \delta_A(q_A, \sigma) \times q_B, \quad \sigma \in \Sigma_A - \Sigma_B \\ q_A \times \delta_B(q_B, \sigma), \quad \sigma \in \Sigma_B - \Sigma_A \end{cases}$$

and : $q_A \in Q_A, q_B \in Q_B$.

Definition 2.2. Run - Computation

A run or equivalently computation of an automaton A on an input sequence $u = \sigma_1, \sigma_2, \ldots, \sigma_n$ is a unique sequence $y = s_1, s_2, \ldots, s_{n+1}$ of states such that $s_1 = i_A$ and for all $j \leq n, \delta(\sigma_j) = s_{j+1}$.

This definition describes the automaton's input reading and the output generation. Note that the first output state is the initial state and hence the output sequence is longer than the input.

2.2. Hybrid Automata

Hybrid automata have been proposed as a formal model for hybrid systems and are a hyper-set of finite state automata. Hybrid automata have in addition, continuous dynamics corresponding to each discrete state of the automaton. These dynamics, as shown above, are typically modeled via differential equations. Figure 2.2 shows a typical example of a hybrid automaton, which models a house thermostat.

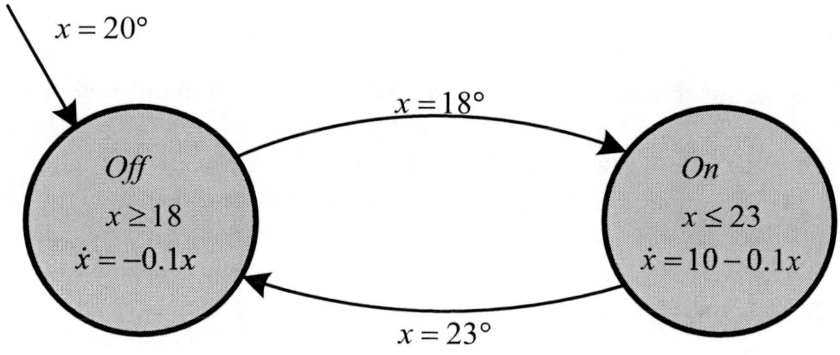

Figure 2.2. A thermostat's hybrid automaton model.

The real valued variable x represents the temperature. The two states are labelled as *On* and *Off* and each one has a flow condition according to which the temperature rises or falls. Initially the temperature is 20 degrees and the heater is off. According to the jump condition the temperature falls, governed by the differential equation $\dot{x} = -0.1 \cdot x$, until the threshold of 18 degrees. Then the heater turns on and the temperature rises according to the flow condition $\dot{x} = 10 - 0.1 \cdot x$, until the upper limit of 23 degrees, where the heater returns to off mode. This simple example denotes the interaction of discrete dynamics (modes on-off) and continuous dynamics (different flow conditions for each mode). After this example a simple definition of hybrid automata is given.

Definition 2.3. Hybrid Automata

A hybrid automaton is defined as $H = \langle Q, X, Init, f, D, E, G, R \rangle$ where

- Q is a finite nonempty set of discrete variables,
- X is a finite nonempty set of continuous variables,
- $Init \subseteq Q \times X$ is a set of initial states,
- $f: Q \times X \rightarrow X$ is a vector field,
- $D: Q \rightarrow P(X)$ is a domain,
- $E \subset Q \times Q$ is a set of edges,
- $G: E \rightarrow P(X)$ is a guard condition and
- $R: E \times X \rightarrow P(X)$ is a reset map.

In bibliography one can also find studies in a subset of hybrid automata, the Zeno hybrid automata [18, 19]. The main difference is that Zeno hybrid automata are systems that accept executions with infinite number of transitions in a finite time interval. Problems appear when trying to simulate a Zeno execution, as a large number of transitions exhibit in a short time interval.

2.3. Timed Automata

Timed automata were first introduced in 1994 by Alur and Drill and since then have become one of the most well studied and widely used models for real-time systems. They are an extension of finite automata as each transition is restricted by an additional set of clocks.

Definition 2.4. Timed Automata

Formally a timed automaton is defined as a 5-tuple $A = \langle Q, \Sigma, i, C, E \rangle$ where:

- Q is a finite nonempty set of states,
- Σ is a finite nonempty set of events, called alphabet,
- $i_A \in Q_A$ is the initial state,
- C is a finite set of clocks and
- $E \subseteq Q \times Q \times \Sigma \times 2^C \times \Phi(C)$ gives the set of the transitions, where $\Phi(C)$ is defined as a set of clock constraints.

An edge $(q, q', a, \lambda, \varphi)$ represents a transition from state q to state q' on input signal a. The set $\lambda \subseteq C$ gives the clocks to be reset with this transition and φ is a clock constraint over C.

2.4. PLC Automata

A very useful tool for the description of distributed real-time systems that are implemented on a Programmable Logic Controller is a PLC automaton. PLC automata were first introduced in 1997 by Henning Dierks and are a subclass of timed automata. The main difference from them is that PLC automata are being restricted to only one clock and certain types of restraints on it.

Definition 2.5. PLC Automata

A PLC automaton is described by a 9-tuple $A = \langle Q, \Sigma, q_0, \delta, \varepsilon, S_e, S_t, \Omega, \omega \rangle$ where:

- Q is a finite nonempty set of states,
- Σ is a finite nonempty set of inputs,
- q_0 is the initial state,
- $\delta: Q \times \Sigma \rightarrow Q$ is the transition function,
- $\varepsilon > 0$ is the upper bound for a cycle,
- $S_e: Q \rightarrow P(\Sigma)\backslash\{0\}$ is a function assigning to each state a set of delayed inputs,
- $S_t: Q \rightarrow \Re_{\geq 0}$ is a function assigning to each state a delay time, that is how long the delayed inputs should be ignored,
- Ω is a finite nonempty set of outputs, and
- $\omega: Q \rightarrow \Omega$ is the output function.

An example of a PLC automaton is shown in Figure 2.3. The automaton has three states $\{q_o, q_1, q_2\}$, two inputs $\{i_1, i_2\}$ and q_o as initial state. The notation in each state denotes the delay time and the corresponding delayed inputs. For example, at state q_1 the input i_1 will be delayed for 4 time units while i_2 will not, and at state q_o both inputs will be delayed for 3 time units.

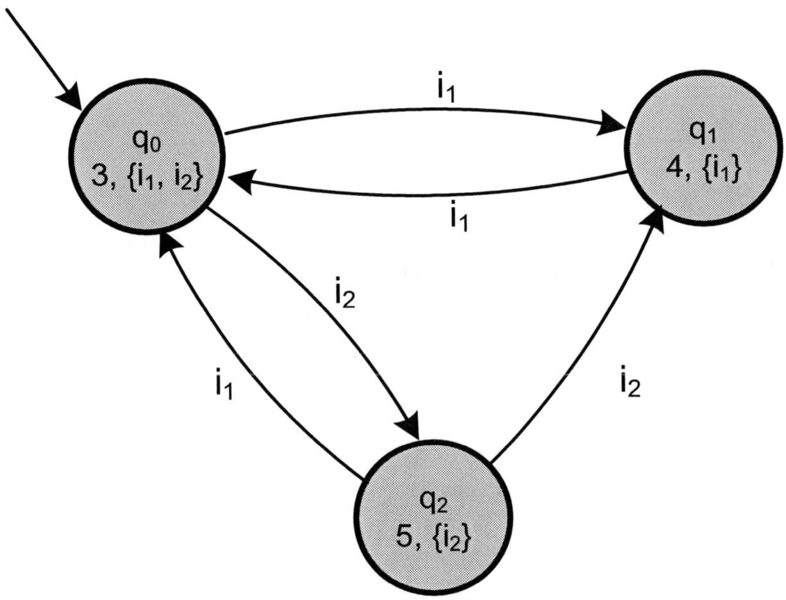

Figure 2.3. A PLC automaton example with three states.

2.5. Software for Modeling Discrete Event Systems

In this section the more essential software tools that have developed for modeling discrete event systems are presented. All of them were developed by research institutes and Universities and are free to public for non-profit use.

2.5.1. JGrafchart

JGrafchart [49] is an implementation of Grafchart, a graphical programming language for sequential control applications, written in Java. Grafchart is based on Grafcet or Sequential Function Charts (SFC), Petri nets, State charts, high level programming languages and object-oriented programming. JGrafchart consists of two parts:

- the graphical editor and
- the run-time system.

A user first creates Grafchart function charts and then compiles and executes them with the run-time system. This feature of simulation and verification is the great advance of JGrafchart, especially for Grafcet-SFC, which is specified in standard IEC 31131-3 as one of the languages for PLC programming.

2.5.2. UPPAAL

UPPAAL [50] is a software tool for modeling, simulation and verification of real-time systems. The tool is suitable for systems that can be modeled as a collection of non-deterministic processes with finite control structure and real-valued clocks, communicating each other through channels or shared variables. UPPAAL consists of three main parts:

- a description tool,
- a simulation tool and
- a model checker.

The description tool describes system behavior as networks of automata extended with clock and data variables. The simulation component is a validation tool, which enables examination of possible dynamic executions of a system. The model checker is used for automatic verification of safety properties by reachability analysis of the symbolic state-space of a system. UPPAAL's typical application areas include real-time controllers and communication protocols in particular, those where timing aspects are critical.

2.5.3. Diagen

DiaGen [51] is a system for easy developing of powerful diagram editors. Some examples of editors, which have been generated with DiaGen are a ladder diagram editor, an editor for hierarchical Petri nets, a state chart editor, flow chart editor etc. These editors can also be used as simulation-verification tools.

2.5.4. Kronos

KRONOS [52] is a software verification tool for real-time systems. KRONOS is based on the theory of timed automata and timed temporal logics. Specifically, the components of real-time systems are modelled as timed automata and the correctness requirements are expressed in the real-time temporal logic TCTL. Some of the main features provided by KRONOS are:

forward and backward analysis, optimisation of the number of clocks used in the model, generation of counter-examples, and binary decision diagrams.

2.5.5. Hytech

HyTech [53] is an automatic tool for the analysis of embedded systems. The system components are described as linear hybrid automata, that is hybrid automata whose continuous dynamics is governed by linear constraints on the variables and their derivatives. HyTech is used to compute the condition under which the system satisfies a temporal requirement. Temporal requirements are verified by symbolic model checking and if the verification fails, HyTech generates a diagnostic error trace.

2.5.6. Moby/PLC

Moby/PLC [54] supports the development of distributed real-time control programs for PLCs. First, the system is modeled as a network of PLC-automata by the central Moby/PLC's component, a graphical editor. Then system can be processed by other components, such as, simulator for networks of PLC-automata, verification components and finally a compiler into programming language ST (Structured Text) for PLCs.

2.5.7. Shift

Shift [55, 56] is a programming language for describing dynamic networks of hybrid automata. Such systems consist of components, which can be created, interconnected and destroyed as the system evolves. Components exhibit hybrid behavior, consisting of continuous-time phases separated by discrete-event transitions. Components may evolve independently, or they may interact through their inputs, outputs and exported events. The interaction network itself may evolve.

2.5.8. Supremica

Supremica [57] is a tool for formal verification and synthesis of control systems. Supremica can be used building a model of the uncontrolled system and the desirable specifications of how the closed-loop system is supposed to behave. Then the software tool generates a suitable supervisor such as to fulfill the specifications. Finally Supremica is able to generate executable code that can be executed by any IEC61131 compatible PLC.

3. Global Automata

Global automata are a generalized formal method for modelling industrial production systems. They are based on automata philosophy bridging the gap between academic methods and real application systems, and are suitable for modelling a large variety of industrial systems. Global automata introduce additional formulation parameters to those of the conventional methods.

3.1. Global Automata Definition

Definition 3.1. Global Automata: A Global Automaton is Defined by the Tuple

$A = \langle X, Z, Q, \Sigma, Init, Flow, L, S, W, E, R_X, R_Z \rangle$, where:

- System's variables:
  - Real-valued variables: $X = \{x_1, x_2, x_3, \dots, x_m\}$
  - Discrete or binary variables: $Z = \{z_1, z_2, z_3, \dots, z_k\}$
- Set of states or locations: $Q = \{q_1, q_2, q_3, \dots, q_n\}$
- Alphabet or set of events: $\Sigma = \{\sigma_1, \sigma_2, \sigma_3, \dots, \sigma_\lambda\}$, which can be:
  - Discrete variables.
  - Conditions over the real-valued variables.
  - Any combination of these.
- Initial conditions for every system variable including the initial state which is indicated with an arrow: *Init*
  - $X = X_0$
  - $Z = Z_0$
  - q_0
- Flow conditions which typically are differential equations for the real variables and logical expressions for the discrete:
  - $F(X, \dot{X}) = 0$
  - $Z_{i+1} = G(Z_i)$
- Invariant conditions: $L = \{l_1, l_2, l_3, \dots, l_n\}$
- Restrictions or safe values: $S = \{s_1, s_2, s_3, \dots, s_n\}$
- The set of events to be ignored until the satisfaction of restrictions: $W = \{w_1, w_2, w_3, \dots, w_n\}$ with $w_i \subseteq \Sigma$.
- Set of transitions: $E \subseteq Q \times Q \times \Sigma \times R_X \times R_Z$

- Reset table for each transition concerning all or some of the system variables:
  - $X = R_X$
  - $Z = R_Z$

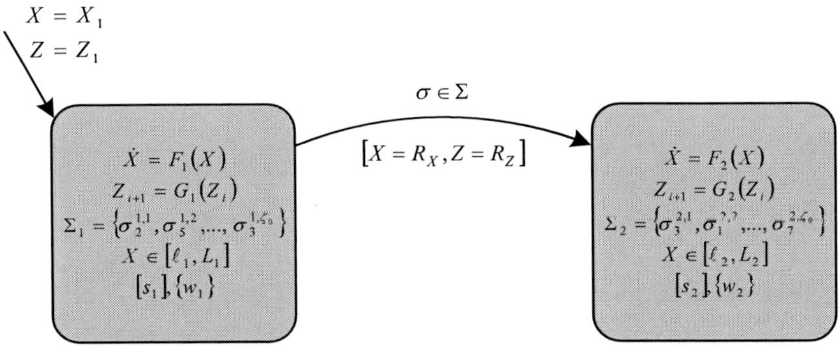

Figure 3.1. A simple global automaton model with two states.

Each set $(q, q', \sigma, r_X, r_Z)$ represents a transition from state q to state q', which is caused by the event $\sigma \in \Sigma$. Set $r_X \subseteq R_X$ gives the real-valued variables to be reset during this transition, while set $r_Z \subseteq R_Z$ gives the discrete variables.

As shown in figure 3.1 each state q_i has a corresponding set of parameters, which are:

- Flow conditions:
  - $F_i(X, \dot{X}) = 0$
  - $Z_{j+1} = G_i(Z_j)$
- Active events at present state: $\Sigma_i \subseteq \Sigma$. Set Σ_i has, by definition, ζ elements, each one of which belongs to set Σ. $\Sigma_i = \{\sigma_j^{i,k}\}$, where i is the present state, $k = 1,2, \ldots, \zeta$ and $j \in [1, \lambda]$. Index k also denotes transitions priority caused by different events. If two events occur simultaneously and cause two different transitions, transition with the lower index k will take place.
- Invariant conditions: l_i
- Restrictions or safe values: S_i
- Set of events to be ignored until the satisfaction of restrictions: $w_i \subseteq \Sigma$

The above definition gives all the specifications needed to construct a global automaton. It is proper to give some more details about this definition to avoid misunderstandings. Global automata are based on automata theory and hence their main feature is the control graph with a number of states and transitions between those states. The definition of states in automata theory implies that a state of a system is the set of all discrete and continuous variables, which is all the knowledge needed to calculate the future behaviour of the system given an input sequence. But in global automata state takes a more general description defined more than operation mode or location, putting aside its former definition. So in this paper, this generalized form of states is used under the title *state*.

In addition, some aspects concerning the control task have to be pinpointed. Firstly, the ability of continuous control can be included in the set of flow conditions in a state. In example, the most common control algorithm in industrial use today, PID controller, can take its common form $u(t) = K_p \cdot e(t) + K_i \int_0^t e(t)dt + K_d \frac{de(t)}{dt}$ using a flow condition.

3.2. Global Automata Structural Properties.

In this section, the structural properties of Global Automata are presented [58].

Definition 3.2. A Global Automaton GA is Linear if

1. all continuous variables of GA take values in linear spaces,
2. for all states $q_i \in Q$, the flow conditions F are all convex,
3. for all transition $T_i \in E$, the jump condition $T_i \in E$ is convex.

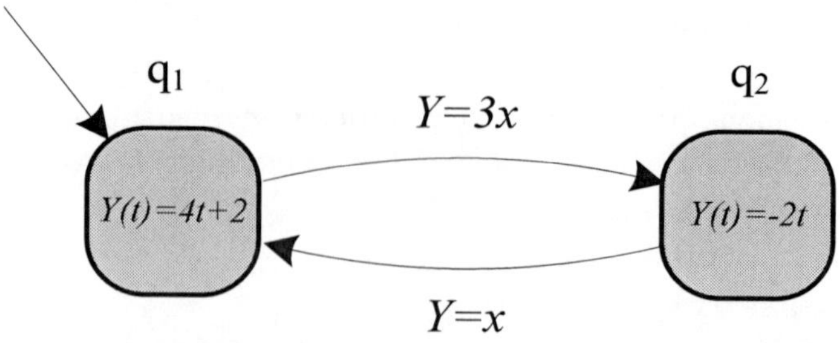

Figure 3.2. Example of a linear automaton.

Linearity in the sense of system theory. According to the formal syntax of GA the flow conditions are generally considered as differential equations. More precisely, the flow conditions can be modeled by a m-dimensional first-order differential equation

$$\dot{x} = f(t, x, u)$$

where x is the state vector in the sense of state space – must not be confused with states-locations of GA – taking values in sum subset of Euclidean space and $f(t, x, u)$ is a control vector field. Since in control theory linear systems dynamics are modeled as $\dot{x}(t) = A(t)x(t) + B(t)u(t)$, the linear flow conditions can be modeled as

$$f(t, x, u) = A(t)x(t) + B(t)u(t)$$

where the state vector $x \in X \subseteq \Re^m$, and the control input vector $u \in U \subseteq \Re^p$, *A(t)* and *B(t)* are matrices in $\Re^m \times \Re^m$ and $\Re^p \times \Re^p$ respectively. The size *m* is called the automaton dimension and the size *p* is the control input size.

Definition 3.3. A Global Automaton GA is Time Invariant if

1. for all locations $q_i \in Q$, the flow condition $Flow(q_i)$ is time invariant, and
2. for all transitions $T_i \in E$, the jump condition $T_i(\sigma_i)$ is time invariant.

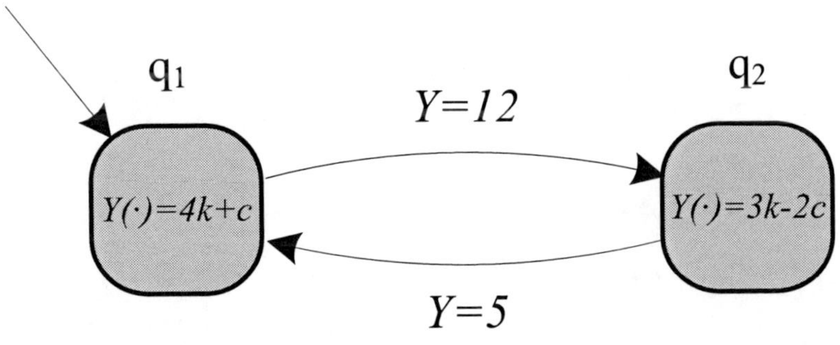

Figure 3.3. Example of time invariant automaton.

In the given example, equation $Y(t) = 4k + c$ denotes that state q_I has time invariant flow conditions, where *k, c* may be constants or time invariant arithmetic values of variables, i.e. a variable denoting tank level. Similarly, all automaton transitions are triggered by time invariant events, as is *Y=12*.

Definition 3.4.

A transition system $T = (Q, \Sigma, \rightarrow, Q_S, Q_F, R)$ consists of the sets Q and R where $R = R_X \cup R_Z$ the alphabet Σ, the initial condition Init(Q) denoted as Q_S, a set $Q_F \subseteq Q$ of final states and a transition relation $\rightarrow$ between them with $\rightarrow \subseteq Q \times Q \times \Sigma \times R_X \times R_Z$.

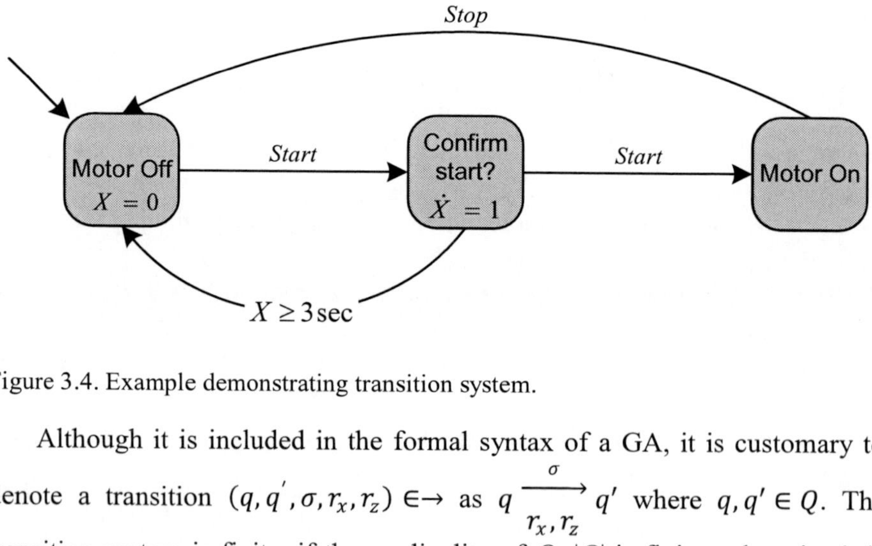

Figure 3.4. Example demonstrating transition system.

Although it is included in the formal syntax of a GA, it is customary to denote a transition $(q, q', \sigma, r_x, r_z) \in \rightarrow$ as $q \xrightarrow[r_x, r_z]{\sigma} q'$ where $q, q' \in Q$. The transition system is finite, if the cardinality of Q, $|Q|$ is finite, otherwise it is infinite.

Definition 3.5.

A transition system $T = (Q, \Sigma, \rightarrow, Q_S, Q_F, R)$ is deadlock-free if for any state $q \in Q$ there exist a state $q' \in Q$ and an event $\sigma \in \Sigma$ such that $q \xrightarrow[r_x, r_z]{\sigma} q'$.

Definition 3.6.

A trajectory of a transition system is finite or infinite sequence of states $\{q_i\}_{i=0}^{N}$ such that

1. $q_0 \in Q_s$ and
2. $q_i \xrightarrow[r_x, r_z]{\sigma} q_{i+1}$ for all $i = 0,1,\dots,N-1$

For the automaton of Fig.6, a possible trajectory may be

$$Motor\ off \xrightarrow{Start} Confirm\ Start \xrightarrow{Start} Motor\ On \xrightarrow{Stop} Motor\ Off$$

Definition 3.7.

Given a transition system $T = (Q, \Sigma, \rightarrow, Q_S, Q_F, R)$ a state $q' \in Q_F$ is reachable from state $q \in Q_S$, denoted by $q \xrightarrow[\{r\}]{\{\sigma\}} q'$, if there is a sequence of transitions for which $q \xrightarrow[\{r\}]{\{\sigma\}} q'$ is true.

The reachability problem consists actually of checking if

$$Post(Q_S) \cap Q_F \neq 0 \text{ or } Q_S \cap Pre(Q_F) \neq 0$$

where for a region P being a subset of states $P \subseteq Q$,

$$Post(P) = \left\{ q \in Q \,\middle|\, \exists \sigma \in \Sigma \ \ \exists p \in P \ \ p \xrightarrow[r_x, r_z]{\sigma} q \right\} \text{ and}$$

$$Pre(P) = \left\{ q \in Q \,\middle|\, \exists \sigma \in \Sigma \ \ \exists p \in P \ \ q \xrightarrow[r_x, r_z]{\sigma} p \right\}$$

In general, a temporal logic formula can be used in order to formulate a global reachability problem. Given a transition system T and a temporal logic formula φ, the basic verification problem is $T| = \varphi$ where φ is specified by using one proper temporal logic.

The verification task is also defined as a searching procedure looking for the satisfaction of a desirable property, expressed by equation φ. The most common properties, known as Safety and Liveness sequence properties, concern the reachability behavior of the automaton. For example, the property

"The state q always remains in a set of states called region $P \subseteq Q$"
is one such property, and so is the dual property
"The state q eventually reaches a set of states called region $P \subseteq Q$"

The first property is known as a safety property because the region P express the "good" or "safe" states. For example, in the extruder system shown in Figure 3.5 and the relevant automaton of Figure 3.6, where we need to ensure that the temperature increment, because of the pressing action on the aluminum billet, will not overcome the already overheating temperature t_h, region P can be the set of states for which $t_c \leq t_h$. We would like to ensure that the incandesced aluminum billet will not pass the melting point (i.e. $t_c \leq t_h$).

Using notation from Temporal Logic this safety property can be written as $\square(q \in P)$ where $\square$ stands for "always". The second property, "the state q reaches p" is known as a liveness property. It reflects the fact that something good should eventually happen to the system. For example, we would like the incandesced aluminum billet not only to avoid melting, but also to be transformed in an acceptable aluminum profile. In the temporal logic notation this property can be written as$\lozenge$ $(q \in P)$ where $\lozenge$stands for "eventually". Using temporal logic, one can encode complex properties of the system such as $\square \lozenge (q \in P)$which stands for "always, eventually q is in P", or in other words, the state visits the set P "infinitely often". Similarly, $\lozenge \square(q \in P)$ stands for "eventually always q is in P" or in other words, t_c satisfies $t_c \leq t_h$ at some point and stays ever after.

According to the definition of the Global Automata, a state $q \in Q$ of GA includes the continuous evolution in time of real-valued variables $x \in X$ as well as of discrette values $z \in Z$, where the first one is caused by the control input $u \in U$. We refer to pair $(q, u) \in Q \times U$ as the powered state of GA. The set U of control inputs is a subset of system's variables.

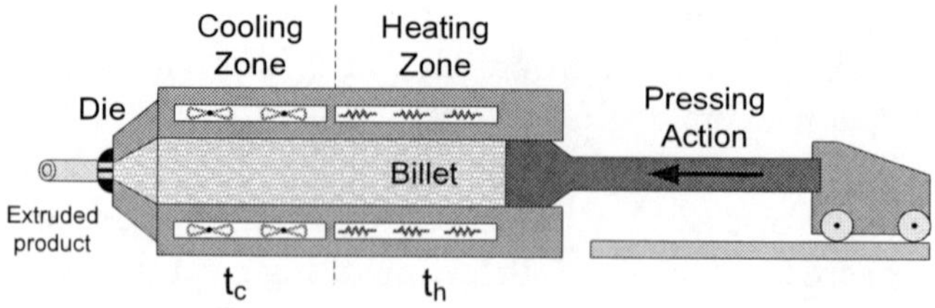

Figure 3.5. Aluminum extruder.

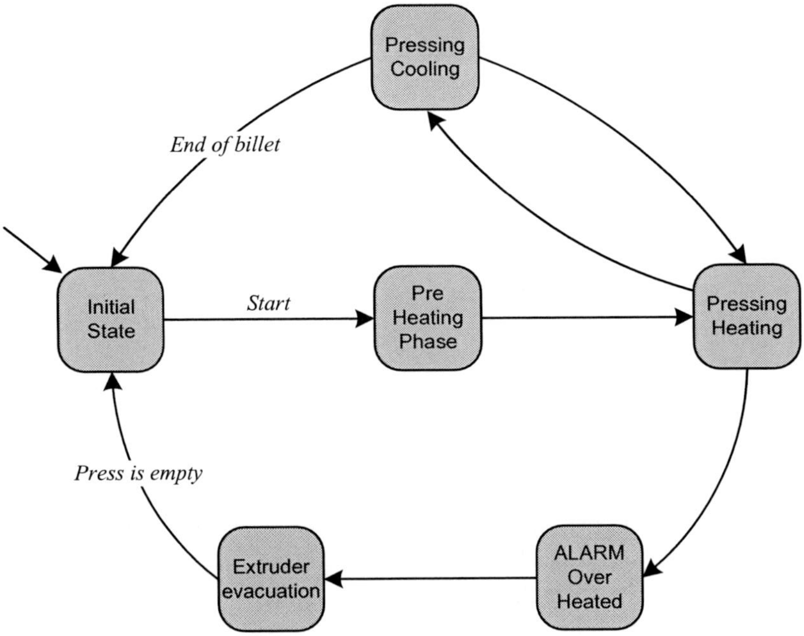

Figure 3.6. Temperature control in aluminum extruder.

Definition 3.8.

Reachability in the sense of control mode. The reachability problem for a global automaton asks, given an GA and a control input vector $u \in U$, if there is an initialized trajectory $\{q_i\}_{i=0}^{N}$ of a transition system T that visits a powered state (q_N, u).

A run or execution of an automaton is a trajectory where all the transitional states are reachable. Note that global automata are deterministic because of event hierarchy introduced in their formal definition. Concerning the extruder example a desired execution of the automaton will be any trajectory excluding the alarm states.

Simulation and bisimulation duality. The composition of two automata help us to model a whole plant – that is a hybrid system - behavior through a supervisor controller so as to prevent any undesirable response by dynamically disabling certain controllable events. The behavioral equivalence between the initial two automata and the composed one can be based on simulation and bisimulation relation.

Definition 3.9.

Given two transition systems $T = (Q, \Sigma, \rightarrow, Q_S, Q_F, R)$ and $T' = (Q', \Sigma', \rightarrow', Q'_S, Q'_F, R')$ a binary relation $\tau \subseteq Q \times Q'$ is called a simulation relation if

1. $(q_1 \, \tau \, q'_1) \wedge (q_1 \in Q_S) \Rightarrow (q'_1 \in Q'_S)$
2. $(q_1 \, \tau \, q'_1) \wedge (q_1 \in Q_F) \Rightarrow (q'_1 \in Q'_F)$
3. $(q_1 \, \tau \, q'_1) \wedge ((q_1, \sigma, q_2) \in \rightarrow) \Rightarrow \exists q'_2$ such that $(q_2 \, \tau \, q'_2) \wedge ((q'_1, \sigma, q'_2) \in \rightarrow)$

We write $(q \, \tau \, q')$ to denote that there exists a simulation relation τ and read as q is simulated by q' or q' can simulate q or q' is similar to q. If a simulation relation exists, then we can say that T' is similar to T. The question is if someone can use the same $\tau \subseteq Q \times Q'$ to show that T is similar to T'. If both statements, T' is similar to T and T is similar to T' are true for the same simulation relation τ, then we can say that T and T' are bisimilar. A bisimulation relation is a symmetric simulation relation which is captured by the following definition.

Definition 3.10.

Given two transition systems, $T = (Q, \Sigma, \rightarrow, Q_S, Q_F, R)$ and $T' = (Q', \Sigma', \rightarrow', Q'_S, Q'_F, R')$ a relation $\cong \subseteq Q \times Q'$ is called a bisimulation relation if

1. $(q_1 \cong q'_1) \wedge (q_1 \in Q_S) \Rightarrow (q'_1 \in Q'_S)$
2. $(q_1 \cong q'_1) \wedge (q_1 \in Q_F) \Rightarrow (q'_1 \in Q'_F)$
3. $(q_1 \cong q'_1) \wedge ((q_1, \sigma, q_2) \in \rightarrow) \Rightarrow \exists q'_2$ such that $(q_2 \cong q'_2) \wedge ((q'_1, \sigma, q'_2) \in \rightarrow)$
4. $(q_2 \cong q'_2) \wedge ((q'_1, \sigma, q'_2) \in \rightarrow) \Rightarrow \exists q_1$ such that $(q_1 \cong q'_1) \wedge ((q_1, \sigma, q_2) \in \rightarrow)$

If T and T' are bisimilar transition systems then

$$T |= \varphi \Leftrightarrow T' |= \varphi$$

The last statement is the basic advantage of bisimulation, since checking $T' |= \varphi$ is equivalent to checking $T |= \varphi$, the complexity of the problem is greatly reduced assuming T' is simplier. The expression of simulation and bisimulation relation in terms of GA automata and not on transition systems is an easy expansion of the above definition. The relation $\cong$ between two automata GA_1 and GA_2 is a bisimulation if and only if it is a simulation of GA_1in GA_2, and at the same time its inverse is a simulation of GA_2in GA_1. This means not only that every computation of GA_1 can be matched by an analogous computation of GA_2 and vice versa, but also that this match can be done through just one relation. In other words, a bisimulation is something more than a pair of simulations.

3.3. Comparison with Existing Methods

Global automata were introduced as a formal method for modelling industrial systems. In this section, global automata are compared to existing methods, based on automata theory and not only. Regarding automata theory, the most widely used form is hybrid automata with great academic acceptance and wide range of applications. Concerning other modelling methods, comparison takes place in contrast with CHARON, Ptolemy and Shift.

Hybrid automata [10] have been proposed as a formal model for hybrid systems, dynamical systems with both discrete and continuous components, which interact with each other. Hybrid systems have been used in manufacturing, automotive engine control, chemical processes, communication networks and other safety critical applications. Hybrid automata model hybrid systems corresponding continuous dynamics to each discrete state of the automaton. Continuous dynamics are hybrid's additional element compared to finite automata and are typically modeled via linear, difference or differential equations. A formal definition of hybrid automata can be found in [10, 12, 13, 14].

A close look to hybrid automata parameters would show that are all included in global automata parameters. To illustrate this conclusion a comparison between two types of parameters has to be made. Most of hybrid automata components are exactly the same as for global automata, such as variables, events, and initial, invariant and flow conditions. The control graph has been replaced by the sets of states and transitions. Finally, jump conditions are included at the set of events. Moreover, global automata have an additional group of parameters. First of all, they have the ability to handle discrete variables with initial conditions and different flow conditions at each state. Furthermore, global automata have transition restrictions if the relative criteria are not accomplished. Restrictions take affect only to events, to be ignored, at each state. Finally, global automata have the ability to reset all system's variables at each transition. Concluding comparison with hybrid automata, one can say that they are a sub-class of global automata. Global automata with extra components present more abilities on modelling industrial systems, since every additional parameter gives to process engineer the opportunity to handle even more system categories.

CHARON [59] is a language for modular specification of interacting hybrid systems based on the notions of agent and mode. Agents describe the architecture of systems while modes do the behaviour of systems. The distinguished feature of CHARON is that it supports both architectural and behavioural descriptions of hybrid systems and provides concurrency and hierarchy in a modular way. It also has the scoping rule of variables partitioned into the set of analogue variables and discrete variables. Continuous behaviour can be specified using differential as well as algebraic constraints, and invariants restricting the flow spaces, all of which can be declared at various levels of the hierarchy. The main features of CHARON are sententiously the architectural and behavioural hierarchy and the discrete and continuous variable updates.

CHARON language is a very powerful tool with complex semantics, and industrial world may be not so familiar with them. Of course, CHARON was not introduced with industrial orientation. An analytical comparison of global automata semantics with CHARON would not be practical since the two methods have different orientation or more precisely, different domain of implementation.

Ptolemy (or more accurately Ptolemy II) [60] is a software framework based on Java, developed as part of the Ptolemy project. The Ptolemy project studies modelling, simulation, and design of concurrent, real-time, embedded systems and Ptolemy II includes a growing suite of domains, each of which realizes a model of computation. It also includes a component library, in which most components are domain polymorphic, in that they can operate in several of the domains. Many domains have been implemented with Ptolemy II, such as finite state machines and modal model, wireless systems, continuous-time modelling etc, while other domains are still experimental. Concluding, Ptolemy is more a visual programming language and not a formulation method. Of course it has a wide variety of application domains but the requirements of the industrial world seem to be too complex and the relative domain is not developed yet.

The last programming tool which can be compared with global automata is Shift, an earlier attempt for developing a simulation tool for hybrid systems. Shift [61, 62] is a programming language for describing dynamic networks of hybrid automata. Shift model was introduced with the belief that offers the proper level of abstraction for describing complex applications such as automated highway systems, air traffic control systems, robotic shop floors, coordinated submarines and other systems whose operation cannot be captured easily by conventional models. Smart-AHS [63] is the only project entirely developed using Shift, as a specification, simulation and evaluation framework for modelling, control and evaluation of Automated Highway Systems. Shift seems to be inactive since 1997, when Smart-AHS was finished. Shift, as mentioned above, was an attempt for developing a tool for hybrid systems, but there was not any limitation on application domain and that was the withdrawal of the whole project. Global automata are focused on industrial systems to clear the obstacle of generality.

Finishing the comparison between global automata and other alternative methods let us summarize the conclusions. Since global automata are a hyper-set of hybrid automata, they also subsume finite, timed and PLC automata. Consequently, every system modelled in one of these forms can be easily translated to a global automaton. Besides, global automata have additional parameters that can be proved to be useful tools on modelling complicated systems. Summarizing, global automata are proved to be less application, or more accurate, domain depended, than every other type of automata. The other three modelling methods, CHARON, Ptolemy and Shift, do not have industrial orientation and hence they cannot be compared in detail with global automata.

4. Modeling Tools on Global Automata

The next two paragraphs present, in detail, two very useful tools of Global Automata for modeling industrial systems. First tool is the State Aggregation and is mainly used to avoid the state space expansion. State aggregation is used to merge two or more states and construct a new hyper-state with exact the same properties as the initial stated. Second tool is the Automata Composition, which is used to compose two or more automata into one. It has to be highlighted that in any case these tools do not change any system characteristics and behavior, since the final model has the same behavior as the initial automata. Finally, the ability of constructing automata models in hierarchical structure is presented as an additional modeling tool.

4.1. State Aggregation

Global automata comprise a new formulation method targeting modelling of industrial systems. Contemporary industrial projects contain *tens of thousands* of states composing a huge overall state space. An enormous state space offers nothing, but only confuses the process engineer. Global automata offer the ability of state aggregation, a useful tool to avoid state space expansion. State aggregation is the merging of two or more states in order to construct a new hyper-state. A hyper-state has the same behaviour as all merged states without losing any kind of data, since real system's behaviour and variables' values remain the same. The only loss may be the visual information that a state offers, when the system is at a particular state (i.e. different states indicating a machine's status). But even this loss can be jumped over any time needed, simply by decomposing a hyper-state in order to have the initial states before merging. Transitions between sub-states of a hyper-state are converted to hyper-state's loop transitions, namely transitions whose final and initial states coincide.

The state aggregation strategy diversifies from system to system, but there are some general rules that the process engineer has to follow.

- State aggregation leads to a new hyper-state defined similarly the sub-states.
- Merging of two states into a new hyper-state adds a new binary variable to the variable set Z. In case of merging more than two states, the extra variable is an integer, added to set X. These variables indicate in which of the merged sub-states system is.
- New hyper-set has the same sets of variables with the addition of binary or integer variable denoted above, while all the other sets are properly modified in order to achieve equivalence.
- Transitions between the original states become loop-transitions of the new hyper-state.

A process engineer, who tries to aggregate a large system, has to be very careful analyzing all the transitions between merged states and all the parameters of each one, especially the events to be ignored and the variables' limits. Concluding, state aggregation is an intuitive procedure where the user plays an integral role. An extensive example of state aggregation is shown in Section 4, where state aggregation is implemented in several steps to achieve a better result, and results in an, as small as possible, state space. In some cases, an expert engineer could end up in a single state automaton, especially useful in case of composition.

For a simple example of state aggregation, consider a machine which starts and stops manually through a pair of start-stop buttons. A model for this system would have two states expressing machines status, Figure 4.1. Merging these two states is simple enough leading to a single state automaton appears, Figure 4.2.

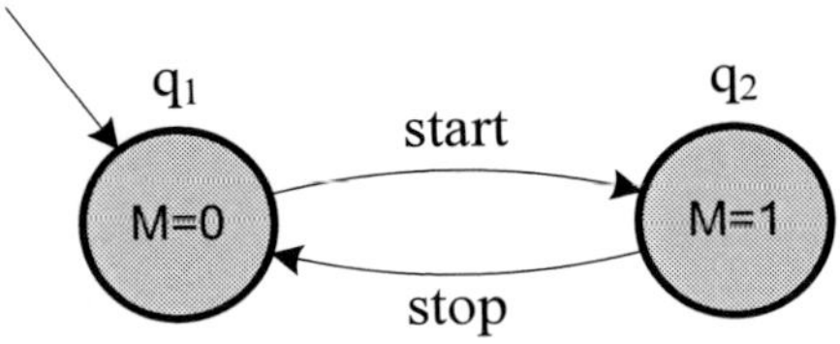

Figure 4.1. Automaton of machine's start-stop operation.

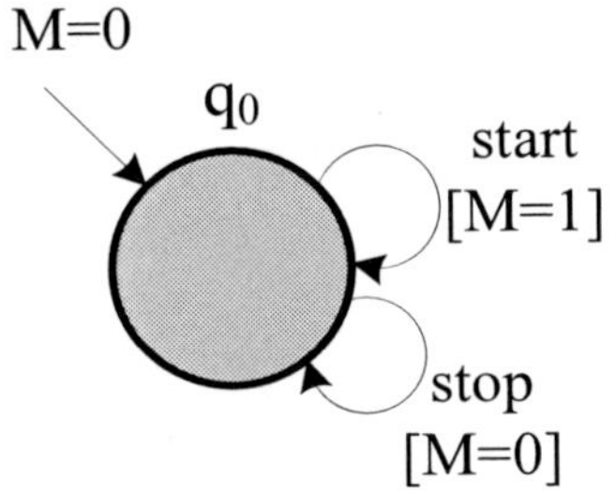

Figure 4.2. Single state automaton.

4.1.1. The "3-Machines Stop" Problem

Let us suppose that three similar machines start and stop manually through an equal number of start-stop buttons. The start-up of the three machines process does not demand any special requirement. Shutdown however, due to operational reasons, takes place immediately for every machine except the last operating one, which must be stopped only if an input signal (e.g. a timer for 30sec) allows it. It is obvious that the other two machines can stop independently of the situation of the input when the corresponding stop button is pressed. Furthermore, the last operated machine is not predefined or constant. On the contrary, it is a stochastic parameter and hence can be any of the three machines. If input has been activated but the stop button of the last operated machine has not been pressed, the machine continues to operate. Summing up, we can claim that any of the three machines can be the last operating one, which must stop in combination with an input while the other two machines will stop in the usual way.

State diagram with all operating combinations of the three machines is shown in Figure 4.3. Let s_i and p_i denote the signals "start" and "stop" of machine i respectively and M_i denotes the operating status of each machine (i = 1, 2, 3). States q_1, q_4 and q_7 are the states where only one machine operates (hence is the last one) and from which the transition to q_0 requires the intermediate state q_6. This state diagram seems similar to common finite state diagrams. But, if we try to take advantage of every capability global automata offer, we can succeed to have state aggregation. The transformation merges states q_1, q_4 and q_7 to a new state labelled *"Only one machine operates"* and states q_2, q_5 and q_8 to a new state, where two machines operate. The new state diagram has only five states, compared with nine at the initial diagram, and is shown at Figure 4.4.

The notation at Figure 4.4 has to be clarified. Firstly, states q_1 and q_2, have six couples of restrictions and events to be ignored. Particularly first three are, $[M_i=1]$, $\{s_i\}$, where $i=1,2,3$ and the latter three are $[M_i=0]$, $\{p_i\}$, where $i=1,2,3$. State q_3, where all machines operate, has no restrictions. All transitions labelled Hs_i are triggered by the events s_i and cause the set of discrete variables M_i to value 1 respectively. Equivalently, transitions Hp_i are triggered by the events p_i and cause the reset of discrete variables M_i to value 0. Finally, transitions Hr_i, which are caused by the events s_i, set discrete variables M_i to value 1 and reset discrete variables M_j (where $j \neq i$) to value 0.

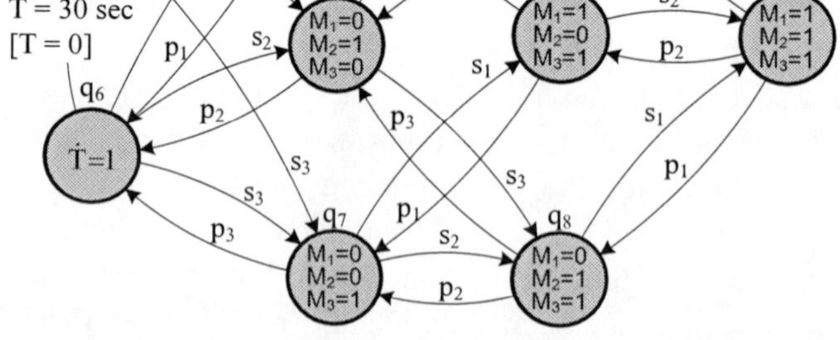

Figure 4.3. Full state diagram for the three machines' stop problem.

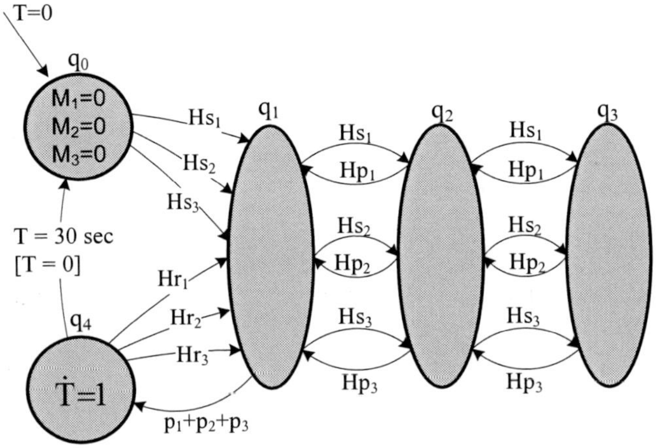

Figure 4.4. Reduced state diagram with five states.

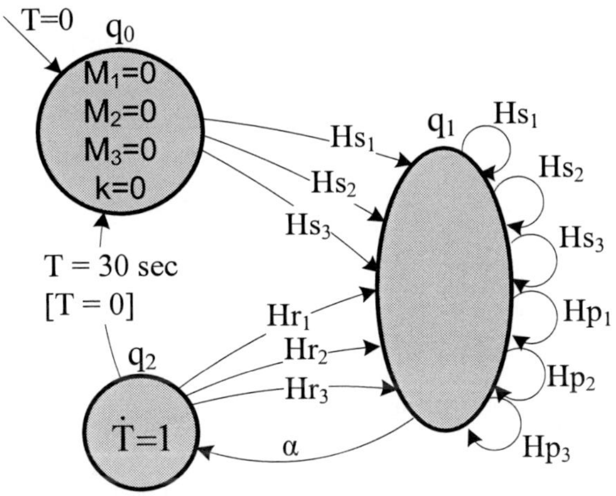

Figure 4.5. Reduced state diagram with three states.

Another approach, adding a new variable k that is the number of operating machines, is shown at Figure 4.5. The new state diagram has only three states, as it merges states q_1, q_2 and q_3 to a new state in which at least one machine operates. State q_1 now has seven couples of restrictions and events to be ignored. The first six are exactly the same as before, $[M_i=1]$, $\{s_i\}$, and $[M_i=0]$, $\{p_i\}$, where $i=1,2,3$, and seventh is $[k=1]$, $\{p_1, p_2, p_3\}$. Similarly to the previous approach, all transitions labelled Hs_i are triggered by the events s_i and cause the set of discrete variables M_i to value 1 and increase counter's k value by one unit. Equivalently, transitions Hp_i are triggered by the events p_i and cause the reset of discrete variables M_i to value 0 and the decrease of counter k. Transitions Hr_i, which are caused by the events s_i, set discrete variables M_i to value 1, reset discrete variables M_j (where $j \neq 1$) to value 0 and initialize counter k to value 1. Finally, event α is defined as $a = M_1 \cdot \overline{M_2} \cdot \overline{M_3} \cdot p_1 + \overline{M_1} \cdot M_2 \cdot \overline{M_3} \cdot p_2 + \overline{M_1} \cdot \overline{M_2} \cdot M_3 \cdot p_3$.

A run or computation of this automaton on an input sequence $u = s_2, s_1, p_1, \alpha, s_3, \alpha, T = 30sec$ is:

$$\left.\begin{array}{c} q_0 \\ \hline M=(0,0,0) \\ T=0 \\ k=0 \end{array}\right| \xrightarrow{s_2} \left.\begin{array}{c} q_1 \\ \hline M=(0,1,0) \\ T=0 \\ k=1 \end{array}\right| \xrightarrow{s_1} \left.\begin{array}{c} q_1 \\ \hline M=(1,1,0) \\ T=0 \\ k=2 \end{array}\right| \xrightarrow{p_1} \left.\begin{array}{c} q_1 \\ \hline M=(0,1,0) \\ T=0 \\ k=1 \end{array}\right| \xrightarrow{\alpha}$$

$$\left.\begin{array}{c} q_2 \\ \hline M=(0,1,0) \\ T=0 \\ k=1 \end{array}\right| \dot{T}=1 \xrightarrow{s_3} \left.\begin{array}{c} q_1 \\ \hline M=(0,0,1) \\ T=0 \\ k=1 \end{array}\right| \xrightarrow{\alpha} \left.\begin{array}{c} q_2 \\ \hline M=(0,0,1) \\ T=0 \\ k=1 \end{array}\right| \dot{T}=1 \xrightarrow{T=30\,\text{sec}} \left.\begin{array}{c} q_0 \\ \hline M=(0,0,0) \\ T=0 \\ k=0 \end{array}\right|$$

The notation:

$$\left.\begin{array}{c} q_0 \\ \hline X = X_0 \\ Z = Z_0 \end{array}\right| \begin{array}{l} \dot{X} = F_0(X) \\ Z_{i+1} = G_0(Z_i) \end{array} \xrightarrow{\sigma} \left.\begin{array}{c} q_1 \\ \hline X = R_X \\ Z = R_Z \end{array}\right| \begin{array}{l} \dot{X} = F_1(X) \\ Z_{i+1} = G_1(Z_i) \end{array}$$

denotes the transition from state q_0 to state q_1 caused by the event σ. The two equations under each state initialize the system variables according to reset table for transition σ. The other two differential equations are current state's flow conditions.

If someone wanted to proceed with aggregation to reduce the model state space, he could add another binary variable. The resulting model has only one state since three states have merged into one with eleven loop transitions, whose initial and final state coincide. This approach is depicted at Figure 4.6.

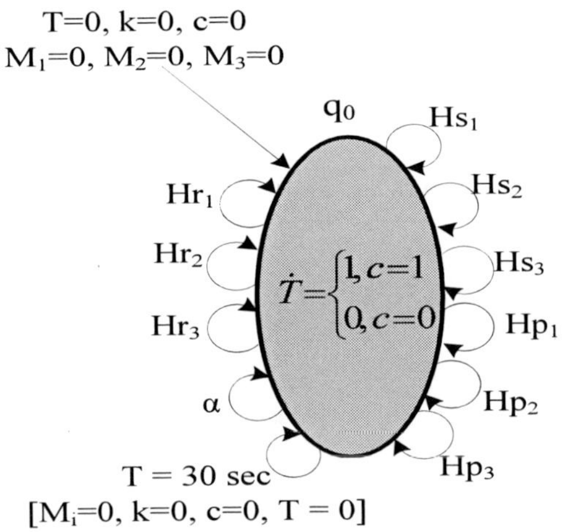

Figure 4.6. Single state automaton for the three machines' stop problem.

State q_0 has seven couples of restrictions and events to be ignored. They all are exactly the same as before, [M_i=*1*], {s_i}, and [M_i=*0*], {p_i}, where *i*=*1,2,3*, and [*k*=*1*], {p_1, p_2, p_3}. Likewise to the prior model, all transitions labelled Hs_i are triggered by the events s_i and cause the set of discrete variables M_i to value 1, increase counter's *k* value by one unit and reset variable *c* to zero. Equivalently, transitions Hp_i are triggered by the events p_i and cause the reset of discrete variables M_i to value 0 and the decrease of counter *k*. Transitions Hr_i, which are caused by the events $s_i \cdot c$, set discrete variables M_i to value 1, reset discrete variables M_j (where $j \neq 1$) to value 0, initialize counter *k* to value 1, reset timer and *c* to zero. Finally, event *α* is defined as $a = M_1 \cdot \overline{M_2} \cdot \overline{M_3} \cdot p_1 + \overline{M_1} \cdot M_2 \cdot \overline{M_3} \cdot p_2 + \overline{M_1} \cdot \overline{M_2} \cdot M_3 \cdot p_3$ and sets discrete variable *c* to 1. Last transition, as shown at the Figure 7, is caused by the event $T = 30sec$ and resets all variables to zero value.

Summarizing the state aggregation process, the conclusion is that for a model which required 9 states initially, state aggregation leads us to a single state automaton. In general, a system with *n* machines would require $2^n + 1$ states. But, by using global automata, a *n*-machines system can be modelled with only three states or even as a single state automaton. The only requirement is an appropriate modification of the last two models presented above. All changes made concern primarily the number of transitions between the states maintaining the type of existing transitions. Another change has been made at event *α*, which now is:

$$a = M_1 \cdot \overline{M_2} \cdot \ldots \cdot \overline{M_n} \cdot p_1 + \overline{M_1} \cdot M_2 \cdot \ldots \cdot \overline{M_n} \cdot p_2 + \cdots + \overline{M_1} \cdot \overline{M_2} \cdot \ldots \cdot M_n \cdot p_n$$

The *n*-machines models are shown below at Figures 4.7 and 4.8.

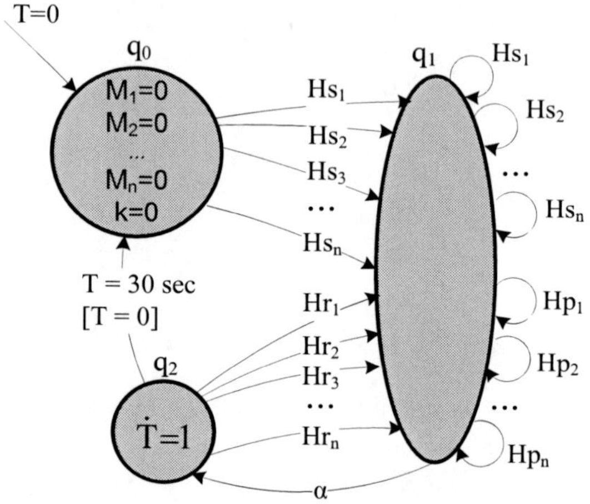

Figure 4.7. Reduced state diagram with three states for the *n*-machines' stop problem.

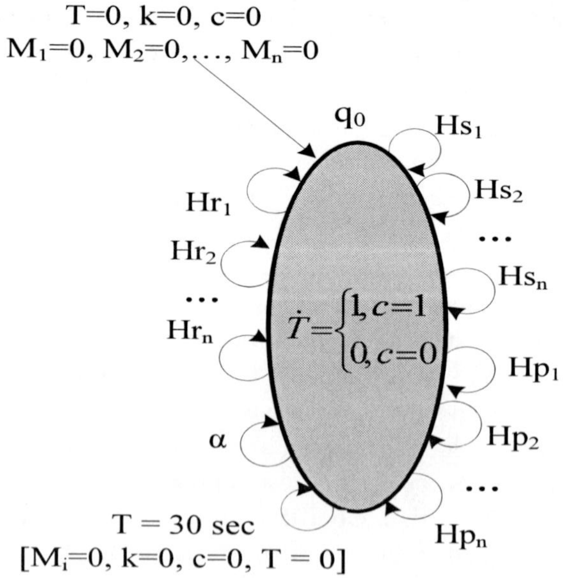

Figure 4.8. Single state automaton for the *n*-machines' stop problem.

4.2. Automata Composition

All academic methods for modeling DESs have a major weakness when considered for industrial use, because they are impracticable for modeling large systems. Present-day industrial production systems have increased complexity and building such system's model is painful. Industrial engineer's work becomes more difficult when the whole system consists of several sub-systems with different characteristics and properties. In such cases, where ad-hoc modeling is not a practical solution, composition can be a useful tool, since it allows the construction of large system model using simple models of all system's components. Small models demand less effort and by putting into practice composition, whole system's model arises. Final model has multiple states compared with primary sub-system models and the summation of all sub-system variables.

Composition is permitted only if sub-systems have different sets of variables and events, which means they are independent. In case of common elements conflict is generated resulting in wrong system modeling. Proper initial system separation to sub-systems implies different sets of variables and parameters in each one.

The composition *A* of two global automata A_1 and A_2 is denoted as $A = A_1 || A_2$. A formal definition of global automata composition is given here.

Definition 4.1. Global Automata Composition:

Consider two global automata $A_1 = \langle X_1, Z_1, Q_1, \Sigma_1, Init_1, Flow_1, L_1, S_1, W_1, E_1, R_{X_1}, R_{Z_1} \rangle$ and $A_2 = \langle X_2, Z_2, Q_2, \Sigma_2, Init_2, Flow_2, L_2, S_2, W_2, E_2, R_{X_2}, R_{Z_2} \rangle$. The composition of A_1 and A_2 is the automaton $A = \langle X, Z, Q, \Sigma, Init, Flow, L, S, W, E, R_X, R_Z \rangle$, where:

- $X = X_1 \cup X_2$
- $Z = Z_1 \cup Z_2$
- $Q = Q_1 \times Q_2$
- $\Sigma = \Sigma_1 \cup \Sigma_2$
- $Init = \left\{ \begin{matrix} [(q_{01}, q_{02})], [(X_{01}, X_{02})], [(Z_{01}, Z_{02})]: \\ (q_{01}, X_{01}, Z_{01}) \in Init_1 \wedge (q_{02}, X_{02}, Z_{02}) \in Init_2 \end{matrix} \right\}$
- $Flow = (F, G)$, where:
  - $F(Q, X) = \begin{bmatrix} F_1(Q_1, X_1) \\ F_2(Q_2, X_2) \end{bmatrix}$
  - $G(Q, Z) = \begin{bmatrix} G_1(Q_1, Z_1) \\ G_2(Q_2, Z_2) \end{bmatrix}$

- $L = \{[(q_1, q_2), (X_1, X_2)]: (q_1, X_1,) \in L_1 \wedge (q_2, X_2) \in L_2\}$
- $S = \{[(q_1, q_2), (X_1, X_2)]: (q_1, X_1,) \in S_1 \wedge (q_2, X_2) \in S_2\}$
- $W = \{[(q_1, q_2), (X_1, X_2)]: (q_1, X_1,) \in W_1 \wedge (q_2, X_2) \in W_2\}$
- $E \subseteq Q \times Q \times \Sigma \times R_X \times R_Z$
- $R_X = R_{X_1} \cup R_{X_2}$
- $R_Z = R_{Z_1} \cup R_{Z_2}$

As noted above, the only condition for composing two automata is their independency.

Lemma 1. Two global automata $A_1 = \langle X_1, Z_1, Q_1, \Sigma_1, Init_1, Flow_1, L_1, S_1, W_1, E_1, R_{X_1}, R_{Z_1} \rangle$ and $A_2 = \langle X_2, Z_2, Q_2, \Sigma_2, Init_2, Flow_2, L_2, S_2, W_2, E_2, R_{X_2}, R_{Z_2} \rangle$, are said to be independent, if and only if, the pairs (X_1, X_2), (Z_1, Z_2) and (Σ_1, Σ_2) are disjoint, that is, $X_1 \cap X_2 = \emptyset$, $Z_1 \cap Z_2 = \emptyset$ and $\Sigma_1 \cap \Sigma_2 = \emptyset$.

4.2.1. A Three Tank System

Consider now an example from a chemical industrial process, which consists of three different tanks, as shown in Figure 4.9. Tanks 1 and 2 contain two different liquid materials and both feed tank 3. The whole procedure starts through a start button. First, valve *g* opens and liquid from tank 1 flows to tank 3. When the liquid volume in tank 3 is 10 lt. valve *g* closes and valve *h* opens. Simultaneously, the mixing process starts by turning on the mixer *p*. Liquid from tank 2 flows until the volume in tank 3 becomes 15 lt. and then, valve *h* closes and the heating phase starts. Heating takes place by turning on heater *k* and stops when the temperature of the mixture rises to 600° C. At that time, mixer and heater turn off and valve *j* opens until tank 3 is empty.

Valve *e* is automatically controlled, so as to keep level in tank 1 between *a* and *b*. When level in tank 1 falls to *b*, valve *e* opens until level arises to *a*. Equivalently, valve *f* is controlled to preserve level in tank 2 between *c* and *d* limits.

The system, described in the previous paragraph, can be seen as a whole, or as composed of three subsystems. Each subsystem models one of the three tanks using a global automaton. The three automata, one for each subsystem, are shown in Figures 4.10, 4.11 and 4.12.

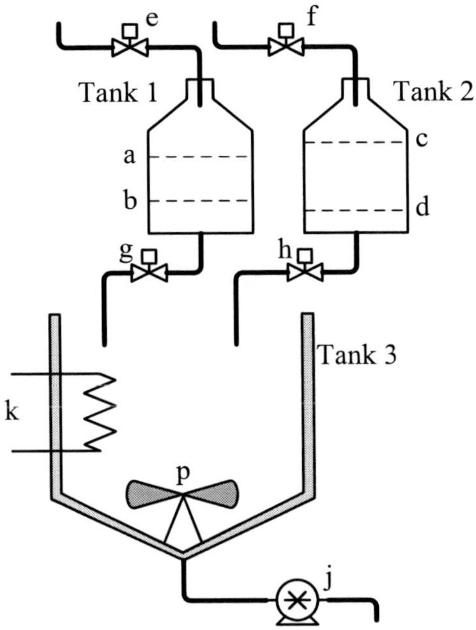

Figure 4.9. Three tank system.

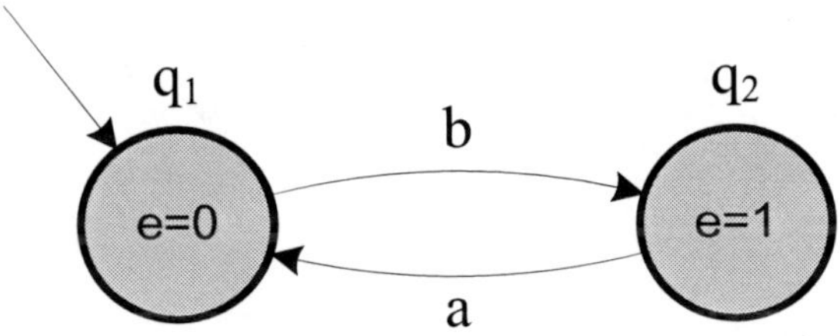

Figure 4.10. First automaton modeling *Tank 1*.

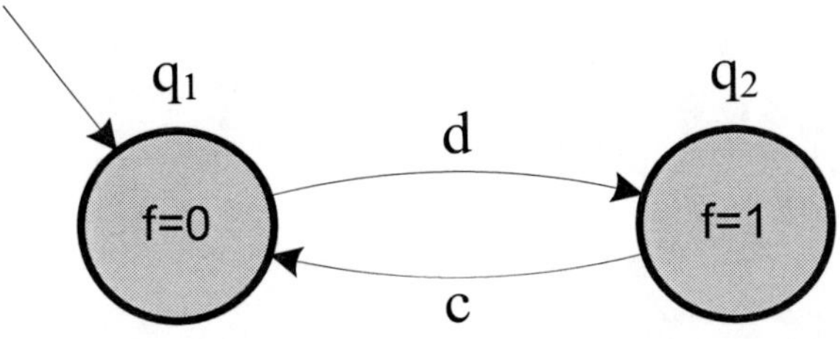

Figure 4.11. automaton modeling *Tank 2*.

First two automata (*automaton 1* and *automaton 2*) are very simple controlling tanks inlet in accordance to liquid's level. Third automaton (*automaton 3*) is slightly more complex. It has a restriction in state q_1, where event s_1 will be delayed until tank 1 is full ($a=1$). In addition, state q_4 has a flow condition according to which temperature arises to the upper limit of 600° C.

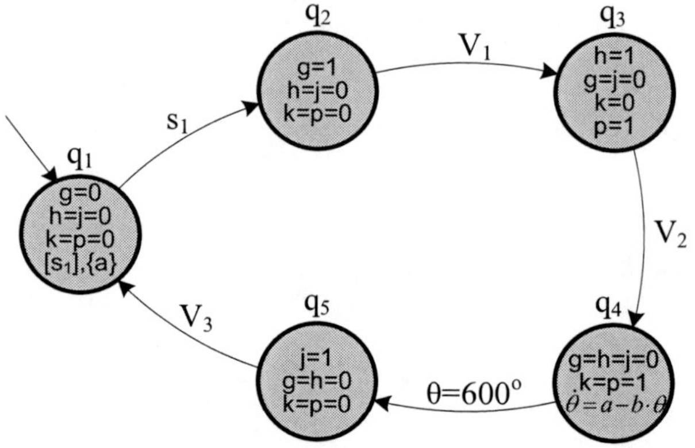

Figure 4.12. Third automaton modeling *Tank 3*.

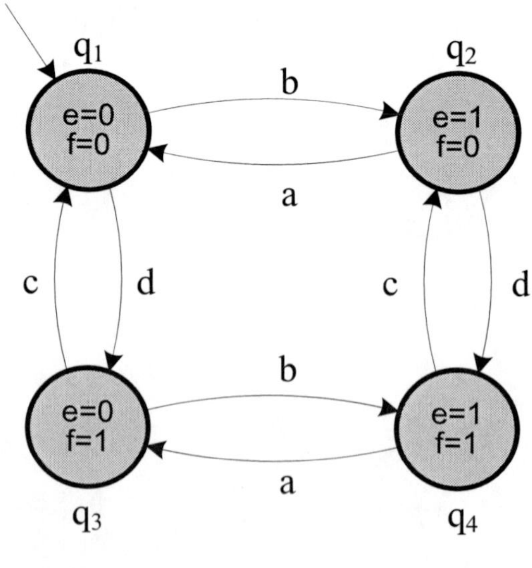

Figure 4.13. Composition of automata 1 and 2.

The three automata shown have to be composed in order to produce the complete system model. Composition, as shown in the previous section, has to be implemented in two stages. First two automata will be composed to one new and then the product will be composed with the third one. According to this, automata of tanks 1 and 2 are composed first. The new automaton (*automaton 12*), which has by definition 4 states and 8 transitions between those states, is depicted in Figure 4.13. *Automaton 3* composed with *automaton 12* gives a new automaton (*automaton 123*) which reconstructs the whole system. *Automaton 123* has 20 states and 60 transitions caused by 9 different events. This automaton is shown in Figure 4.14. States are labelled as Q_{ij}, where $i = 1,2,3,4$ and $j = 1,2,3,4,5$:

The model parameters in each state arc as follows

- Q_{1j}: $e = 0, f = 0$
- Q_{2j}: $e = 1, f = 0$
- Q_{3j}: $e = 1, f = 1$
- Q_{4j}: $e = 0, f = 1$
- Q_{i1}: $g = 0, h = 0, j = 0, k = 0, p = 0 \ /[s_1], \{\alpha\}$
- Q_{i2}: $g = 1, h = 0, j = 0, k = 0, p = 0$
- Q_{i3}: $g = 0, h = 1, j = 0, k = 0, p = 1$
- Q_{i4}: $g = 0, h = 0, j = 0, k = 1, p = 1, \dot{\theta} = a - b \cdot \theta$
- Q_{i5}: $g = 0, h = 0, j = 1, k = 0, p = 0$

The 9 events that cause transitions between states as shown below:

- a: $Q_{2j} \to Q_{1j}$, $Q_{3j} \to Q_{4j}$
- b: $Q_{1j} \to Q_{2j}$, $Q_{4j} \to Q_{3j}$
- c: $Q_{2j} \to Q_{3j}$, $Q_{4j} \to Q_{1j}$
- d: $Q_{1j} \to Q_{4j}$, $Q_{3j} \to Q_{2j}$
- s_1: $Q_{i1} \to Q_{i2}$
- V_1: $Q_{i2} \to Q_{i3}$
- V_2: $Q_{i3} \to Q_{i4}$
- V_3: $Q_{i5} \to Q_{i1}$
- $\theta = 600^o$: $Q_{i4} \to Q_{i5}$

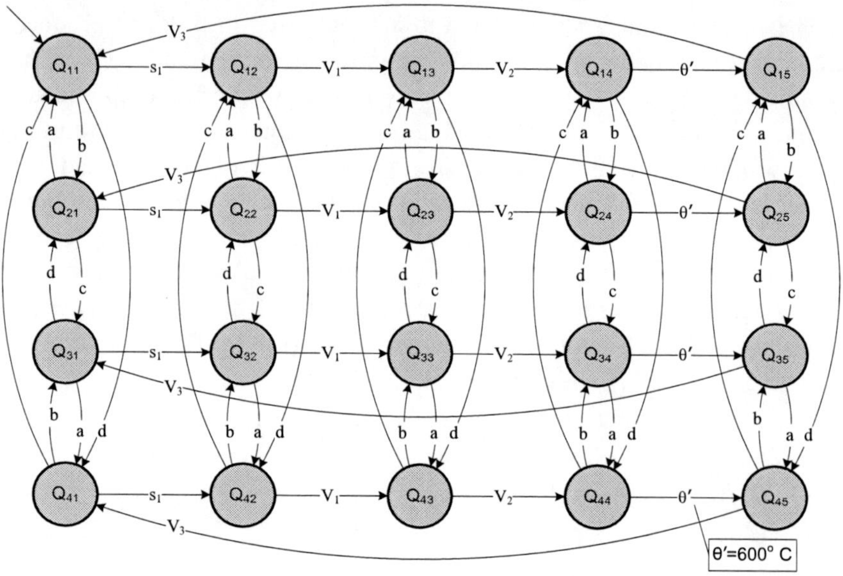

Figure 4.14 *Automaton 123*. Whole system's model.

It must be noted that one may have designed *automaton123* without using the composition of *automata 1, 2* and *3*. Of course this is a much more difficult task compared to the composition described above even for this simple example. For a large industrial system, the composition procedure seems to be the only realistic way of modelling.

In the paragraph where state aggregation was presented, the fact that aggregation leading to a single state automaton is a very useful tool in cases of composition was underlined. The three tank system is a good example, since transformation of the first two automata to a single state, would result in a final model with only five states. This derives from the fact that a composed model has $m \times n$ states, where *m* and *n* are the states of the systems to be composed. In current example, the three automata, without state aggregation, generated a model with $2 \times 2 \times 5 = 20$ states. In case of using state aggregation final model will have only $1 \times 1 \times 5 = 5$ states. Single state automata modelling tanks 1 and 2 are shown in Figure 4.15, while in Figure 4.16 the result of composing them is depicted. Composed automaton, after second stage of composition is depicted in Figure 4.17.

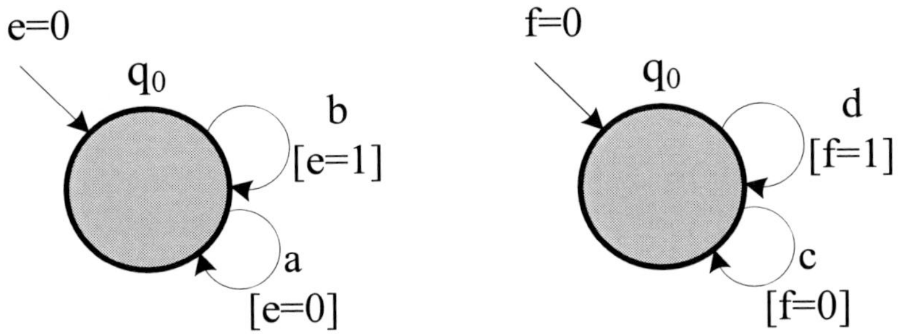

Figure 4.15 Single state automata for *Tanks 1 and 2*.

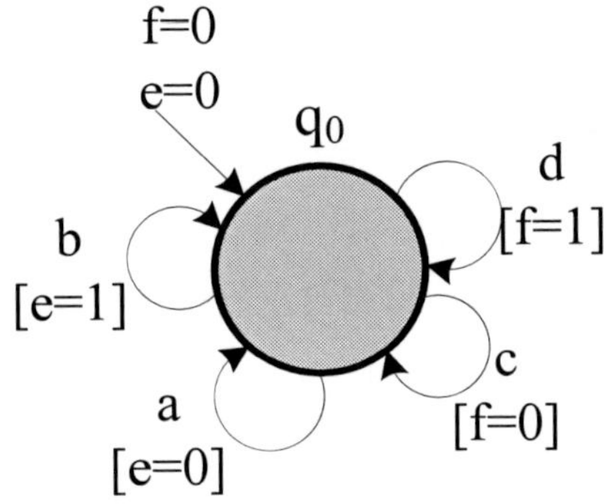

Figure 4.16. First step of composition.

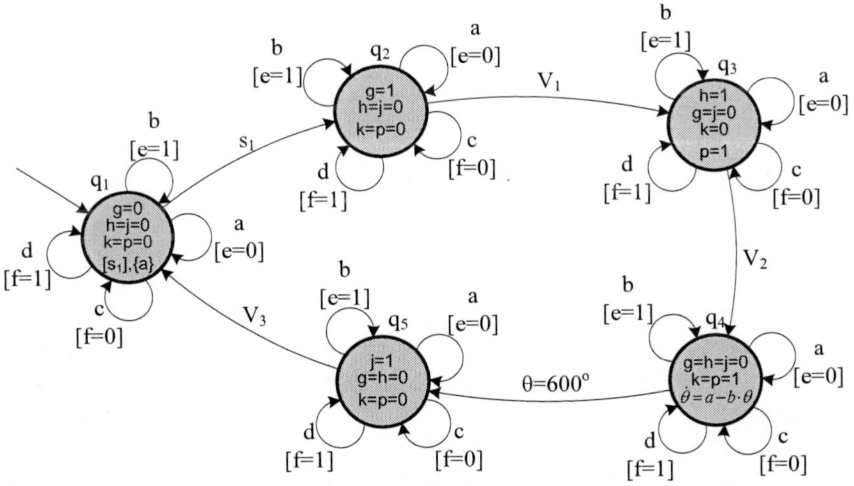

Figure 4.17. Final model with only five states.

4.3. Hierarchy on Global Automata

In order to meet the demands of the contemporary industrial systems there is the necessity of a model with hierarchical structure. Global automata offer the convenience of building such models. In this case, several automata are used each one modeling a specific part of the overall system. Finally, one automaton plays the role of the supervisor controller feeding all the other controllers with specific set-points in order to achieve the desired system behavior. A hierarchy model is not limited in just two layers, since there is the possibility to construct models with more layers. In this case, each automaton receives parameters/inputs from the automaton of the higher layer, while it "feeds" with its outputs the automaton of the lower layer. A detailed example of using hierarchy is given in paragraph 6.2, where a Concrete Batching and Mixing Plant is studied.

5. Implementation Tools

Up to this point, only the modeling tools of Global Automata have been presented, but since the task of a Control Engineer is to control a system, two useful tools for implementing the produced models are introduced. First tool is used mainly to produce executable code for a high level programming language. The produced code can be used either for simulating the model or for controlling the real system if it is introduced in a proper controller with such programming environment. In example, SCADA software with embedded script editors could be a proper implementation platform for such an application. The latter presented tool is used for producing executable code for Programmable Logic Controllers. Synthesis tool has the form of writing scripts in the most common programming languages as they are described in IEC 61131-3 standard. Both tools are presented also via representative examples.

5.1. Guide for Building Simulation Models Based on Global Automata

Advances in technology invariably lead to the construction of systems with additional layers of complexity being wrapped around more primitive but equally complex sub-systems. Simulation provides a means by which such systems, just as real world systems, may be understood and evaluated by duplicating the behaviour of them through software and hardware. Formally, simulation is defined as: *"... the process of designing a model of a real system and conducting experiments with this model for the purpose of understanding the behaviour of the system or of evaluating various strategies (within the limits imposed by a criterion or a set of criteria) for the operation of the system"* . Elucidating this definition one can say that a simulator is a collection of hardware and software systems which are used to mimic the behaviour of some entity or phenomenon. Typically, the simulated entity is from the domain of the tangible, ranging from the operation of integrated circuits to behaviour of complex industrial systems. Simulators may also be used to analyze and verify theoretical models which may be too difficult to grasp from a purely conceptual level. As such, simulators provide a crucial role in both industry and academia.

Despite the increasing recognition of simulators as a viable and necessary research tool, one must also be aware of the possible problems simulators may introduce. Such problems are related whether to the computational limitations of existing hardware platforms, or to the complexity associated with the systems being simulated. The advantages and the disadvantages of simulation are sententiously described below.

One of the primary advantages simulators have is that they are able to provide users with practical feedback when designing real world systems. This allows the designer to determine the correctness and efficiency of a design before the systems are actually constructed, resulting in exploring the merits of alternative designs. The ability of investigating the effects of specific design strategies during the design phase and not during the construction phase reduces the overall cost of building the system. In addition, simulators permit system's designers to study a problem at several different levels of abstraction. A hierarchical approach on a system analysis is especially useful when dealing with the construction of complex systems. Finally, simulators can be used as an effective means for teaching or demonstrating concepts to students. This is particularly true for simulators that take advantage of computer graphics and animation. Such simulation tools show dynamically system's behaviour providing the user with a meaningful understanding of system's nature.

Despite these significant advantages, simulators also have some disadvantages or drawbacks. Many of them can be attributed to the computationally intensive processing required by some simulators. As a consequence, simulation results may not be readily available after the simulation has started. In example, an event that occurs instantaneously in real world may actually take minutes or even hours to mimic in a simulated environment. Such delays are due to complex interactions of systems' entities or even due to physical systems nature. However, as more powerful hardware platforms become available, this problem becomes less of concern. Another way to face up with the aforementioned complexity is to introduce simplifying assumptions or heuristics into the simulator. While this technique can dramatically reduce simulation time, it may also give its users a false sense of security regarding the accuracy of simulation results.

In spite of its weaknesses, simulation is acknowledged as a very useful tool for control engineers. Recognizing the use of simulation, industry, government and academia admonish the presence of simulation courses in universities. Therefore, to meet the existing and on the horizon needs of the scientific and engineering workforce, many universities have curriculum offering graduate degrees in modeling and simulation. In this section, a guide for building simulation models of industrial systems based on global automata is presented.

5.1.1. Guide for Code Generation

Guide, which was introduced in [64], has the form of an algorithm for writing scripts in several programming languages. The steps for building a system's model can be summarized as given below, while an example of the completed form is shown in figure 5.1:

- *Variables declaration.* Firstly, discrete and real-valued variables have to be declared. Additionally, an extra integer variable Q denoting system's state is necessary.
- *Initial conditions.* After declaration, all variables have to be initialized according to automaton's initial conditions. Additionally, variable Q takes the proper value denoting initial state q_0..
- *Execution Loop.* Execution loop is the heart of the program. In this stage takes place the execution and all the helpful results of a system's operation arise.

```
FOR I = 1:1000
    t = t + 0.001
    SWITCH Q
        CASE 1
            Z_1(i+1) = Ā + Z_1(i) · B
            X_1(i+1) = X_1(i) + ...
            X_2(i+1) = f(t)
      IF event_1 occurs AND is not ignored THEN
            Q = 4
            X = R_X1
            Z = R_Z1
    ENDIF
      IF event_2 occurs AND is not ignored THEN
            Q = 3
            X = R_X2
            Z = R_Z2
    ENDIF
        END – CASE
        CASE 2
            ...
        END – CASE
        ...
    END – SWITCH
END – FOR
```

Figure 5.1. Guide's complete form.

If we are working with MATLAB or so, execution loop is a simple FOR-NEXT loop and first of all we have to determine the number of iterations using an additional variable (e.g. FOR I=1:1000). Usually, each simulation's main component is time; therefore each loop has to be assigned to a time value using a command like t=t+0.001 in order to succeed millisecond time accuracy.

Commands representing state's flow conditions. Discrete variables' flow conditions are Boolean equations. Real variables' flow conditions can take the form of time functions or difference equations, which are the analogues of differential equations in the finite difference calculus [65].

Commands representing state's transitions. Each state has a set of active events which trigger a transition. Active events have to be included in state's case structure, taking an IF – THEN formation. Each transition is caused by an event, if it is not ignored, and results in the state change and the reset of some variables. If there are more than one event in a state, relative IF – THEN structures are given in compliance with set Σ_i, which represents hierarchy of active events in *i* state.

5.1.2. Examples of Building Simulation Models by Using The Guide

This section presents some examples illustrating guide's use on building simulation models of industrial systems. These examples are implemented in MATLAB and CX-Supervisor, but software environments are not limited, since everything which provides a script editor can be used. MATLAB is used primarily because is the most common language in Universities all over the world. Regarding CX-Supervisor, it is used because of its nature as a SCADA software. SCADA is the acronym for Supervisory Control And Data Acquisition, a computer for gathering and analyzing real time data used to monitor and control a plant. SCADA systems were first used in the 1960s and since then, most of industrial engineers have become familiar with their use. This is the main reason for choosing CX – Supervisor as the software platform for illustrating the guide use for building simulation models according to global automata, since they were presented with industrial orientation. In addition, CX – Supervisor comes with an embedded Script Editor and a Run-Time environment, where simulation of a system can take place.

5.1.2. A House Thermostat

A global automaton which models a house thermostat is depicted in Figure 5.2. The real valued variable *x* represents the temperature. The two states are labelled as *On* and *Off* and each one has a flow condition according to which the temperature rises or falls. Initially the temperature is 20 degrees and the heater is off. According to the jump condition the temperature falls until the threshold of 18 degrees. Then the heater turns on and the temperature rises until the upper limit of 23 degrees, where the heater returns to off mode.

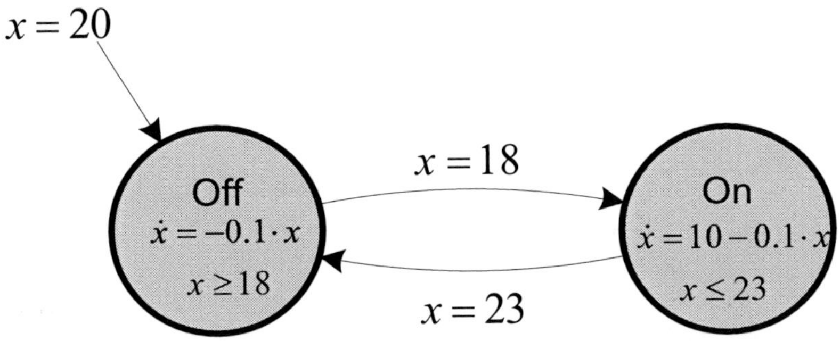

Figure 5.2. A thermostat's global automaton model.

This simple automaton has been implemented in MATLAB. The code written according to guide described in previous section is shown here below, whereas in Figure 5.3 is depicted a plot generated after execution, showing temperature's fluctuation.

```
x(1)=20;          %Initial temperature
k=0;              %Thermostat's initial status
q=1;              %Initial state
for i=1:100
   switch q
      case 1
         x(i+1)=0.9*x(i);          %Difference equation analogue to x'=-0.9x
         k=0;
         if x(i+1)<18
            q=2;
         end
      case 2
         k=1;
         x(i+1)=10+0.9*x(i);   %Equation analogue to x'=10-0.1x
         if x(i+1)>23
            q=1;
         end
   end
end
plot(x),grid
```

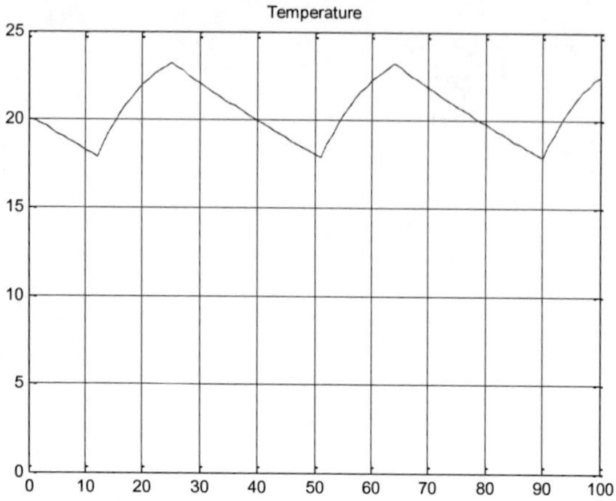

Figure 5.3. Temperature's fluctuation.

5.1.5. *Token Passing Bus Network*

The automaton of figure 5.4 models a local area network with four nodes using the token passing bus protocol. The four states reconstruct the virtual ring between the nodes. When the automaton is in qi state, i node has the token and the capability to use the network. The variable t represents the time and τ is an additional clock variable. The variables M1, M2, M3 and M4 represent the number of queuing packets in each node. The discrete variables s1, s2 s3 and s4 represent the arrival of a new packet at the queuing list in each node. The constant parameter Ts is the token's transmission time, while Ps is a single data packet's transmission time.

All transitions labeled Hs_i are triggered by the events s_i and increase counters M_i by one unit, for *i=1,2,3,4*. In the other hand, transitions Ht_i are triggered by the event $\tau = T_s + P_s$ and cause the reset of variable τ to zero value and decrease counters M_i by one unit, for *i=1,2,3,4*. Finally, transitions HT are caused be the event $\tau = T_s$ and reset variable τ to zero value. In each state are two flow conditions, one for each clock variable: $\dot{t} = 1$ and $\dot{\tau} = 1$. There is also a couple of an ignored event and a restriction, which are: $[\tau = T_s], \{M_i > 0\}$. This means that when the automaton is at q_i state and $M_i > 0$, the event $\tau = T_s$ will be ignored. So, a transition from state q_i to state q_{i+1} takes place after T_s time units, if $M_i = 0$, or else after T_s+P_s time units.

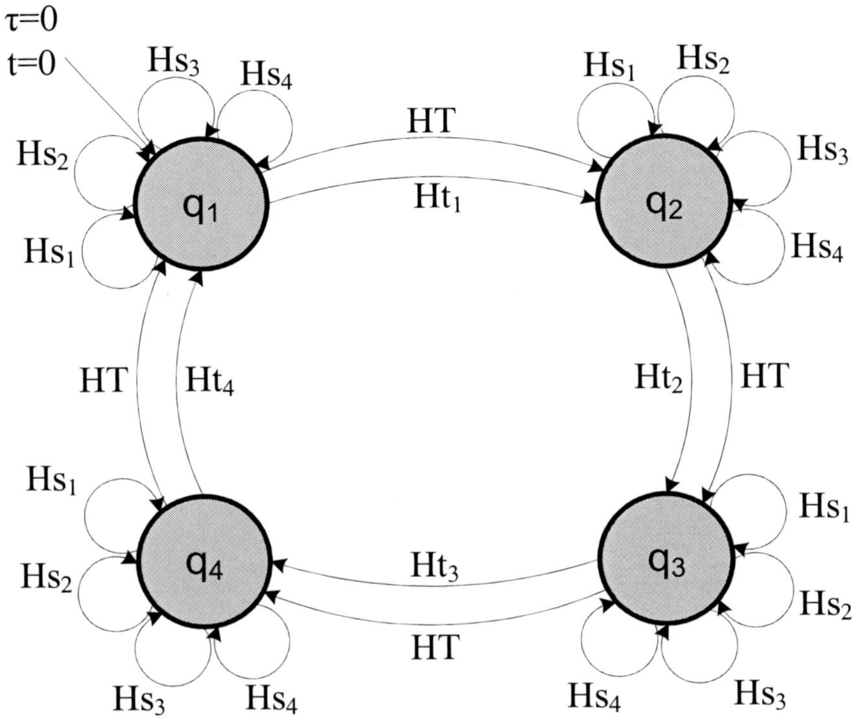

Figure 5.4. Token Passing Bus Model with four nodes.

Initially the automaton is in q_1 state and stays there until $t=T_s$, if $M_1=0$, or until $t=T_s+P_s$, if $M_1>0$, where M_1 is the number of queuing packets at the queuing list of node 1. The above condition means that if node 1 has not any packets to send, the bus's management will pass to node 2 after the token's time (T_s) elapses. In the other case ($M_1>0$), node 1 will send the first packet of his list and the transition to state 2 will take place after T_s+P_s time units (token's time and time for one packet respectively). These two different types of transitions cause the change from one state to the following one.

Each state has also four loop transitions caused by the events s_i, which represent the arrival of a new packet at the queuing list of node i. Consequently, the corresponding counter M_i increases by one unit.

A run of this automaton is shown above presenting the system's states and variables in connection with time. For this example we assumed that $T_s=1$ and $P_s=2$.

$$\begin{array}{c|c} q_1 & \\ \hline M=(1,0,2,0) & \\ t=0 & \dot{t}=1 \\ \tau=0 & \dot{\tau}=1 \end{array} \xrightarrow{\tau=3} \begin{array}{c|c} q_2 & \\ \hline M=(0,0,2,0) & \\ t=3 & \dot{t}=1 \\ \tau=0 & \dot{\tau}=1 \end{array} \xrightarrow{\tau=1}$$

$$\begin{array}{c|c} q_3 & \\ \hline M=(0,0,2,0) & \\ t=4 & \dot{t}=1 \\ \tau=0 & \dot{\tau}=1 \end{array} \xrightarrow{s_2} \begin{array}{c|c} q_3 & \\ \hline M=(0,1,2,0) & \\ t=5 & \dot{t}=1 \\ \tau=1 & \dot{\tau}=1 \end{array} \xrightarrow{\tau=3}$$

$$\begin{array}{c|c} q_4 & \\ \hline M=(0,1,1,0) & \\ t=7 & \dot{t}=1 \\ \tau=0 & \dot{\tau}=1 \end{array} \xrightarrow{\tau=1} \begin{array}{c|c} q_1 & \\ \hline M=(0,1,1,0) & \\ t=8 & \dot{t}=1 \\ \tau=0 & \dot{\tau}=1 \end{array} \xrightarrow{\tau=1}$$

$$\begin{array}{c|c} q_2 & \\ \hline M=(0,1,1,0) & \\ t=9 & \dot{t}=1 \\ \tau=0 & \dot{\tau}=1 \end{array} \xrightarrow{s_1} \begin{array}{c|c} q_2 & \\ \hline M=(1,1,1,0) & \\ t=11 & \dot{t}=1 \\ \tau=2 & \dot{\tau}=1 \end{array} \xrightarrow{\tau=3}$$

5.1.6. The "3-Machines Stop" Problem

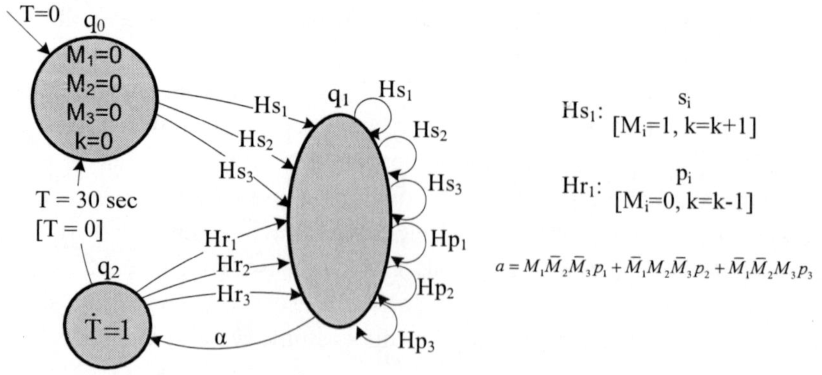

Figure 5.5. Reduced state diagram with three states.

The second system has already presented in paragraph 4.1.1. For the sake of this example, the model with the three states of Figure 5.5 was used.

Code was generated in accordance with MATLAB's notation using the integrated Editor as shown here:

```
case 1
        if s1==TRUE & M1==FALSE
           M1=TRUE;
           k=k+1;
        end
        if s2==TRUE & M2==FALSE
           M2=TRUE;
           k=k+1;
        end
        if s3==TRUE & M3==FALSE
           M3=TRUE;
           k=k+1;
        end
        if p1==TRUE & M1==TRUE & k~=1
           M1=FALSE;
           k=k-1;
        end
        if p2==TRUE & M2==TRUE & k~=1
           M2=FALSE;
           k=k-1;
        end
        if p3==TRUE & M3==TRUE & k~=1
           M3=FALSE;
           k=k-1;
        end
        if a==TRUE
           q=2;
        end
end
```

5.2. Synthesis Tool for Implementing Global Automata in PLCs

Since the 1970s PLC has been the primary workhouse of industrial automation. A PLC is a computer which is suitable for the application control tasks because of its special qualities concerning robustness and response time.

PLCs are characterized by the two different tasks they have to fulfil. First task is to act as an Input/Output device supervising the process, by reading the values of sensors connected to their input ports and setting the values of actuators connected to their output ports. The latter task is applying the control strategy by specifying the relation between the sequence of inputs and outputs which is done by software. The industrial standard approach is the direct implementation of the controller using a PLC programming language. In order to have standardization on programming PLC, the first version of the norm 61131-3 was published by *International Electrotechnical Commission* in 1993 and the current second version in 2003 [46] and since then, it has been adopted by major PLC manufacturers. The international standard defines common elements of programmable controller programs, including program structure, variable declarations etc. and semantics and syntax of several programming languages. Global portability of the code among different PLCs cannot be achieved because the standard is not strict enough. Even though, the program structure is unified.

The IEC 61131-3 standard has a set of five languages for programming PLCs which are Instruction List (IL), Ladder Diagram (LAD), Function Block Diagram (FBD), Structured Text and Sequential Function Charts (SFC). In Germany, the assembler language IL predominates especially under the trademarks of Siemens Step5 and Step7. LAD is based on concepts of control very traditional in the field, and do not reflect recent evolutions in computer science. These two, with FBD language are generally supported by all manufacturing firms. On the contrary, Structured Text and Sequential Function Charts, which are high level languages, are not supported by all PLC types. A number of approaches exist to the formalization of PLC languages. They usually concentrate on the particular language of the norm, rather than on their interaction, and they concentrate on the formal verifications of programs, rather than on other aspects such as code generation.

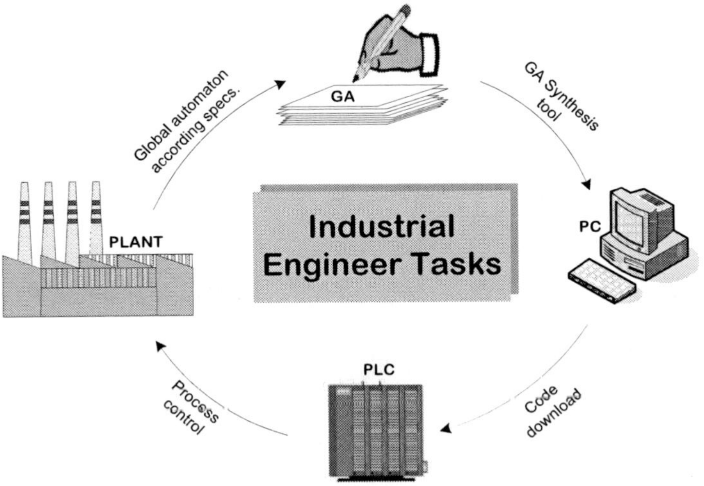

Figure 5.6. Industrial Engineer tasks.

Regarding industrial systems modelling and control with PLCs, it has to be mentioned that, a PLC may be treated as a DES (Discrete Event System), since it changes its state, memory variables and outputs, in response to changes on its inputs and time. Consequently, hybrid systems implemented in PLCs loose their continuity since everything is discretized in accordance to PLC's CPU clock. However, this is not an obstacle on controlling industrial/hybrid systems with PLCs.

In this section is presented a synthesis tool for implementing global automata in PLCs based on the IEC 61131-3 programming norm. This tool describes all necessary steps which an industrial engineer has to follow in order to implement a control strategy, as described in a global automaton, in a PLC. The control graph and the whole set of an automaton parameters are transformed into code which is then downloaded to the PLC to control the system. The tasks of an industrial engineer who has to deal with an industrial plant are shown schematically in Figure 5.6. The plant inspection and recording of all desired specifications are the first and most crucial steps. According to the plant's characteristics and control specifications, the engineer creates the global automaton which meets the process demands. When the global automaton is completed and carefully verified through simulation it has to be translated into PLC code by using the synthesis tool presented here. Finally, the produced code is downloaded to the PLC controller and the whole plant is putting into use for a test running. Possible malfunction comes with reprocessing of plant specifications and revising of the initial automaton. This loop may be essential for accomplishing the goal of controlling the plant.

Synthesis tool has the form of writing scripts in the most common programming languages as they are described in IEC 61131-3 standard. Firstly is given a summary of the required steps and afterwards the analytical description of writing scripts as the standard specifies. The steps for programming are briefly given below.

- *Variables declaration.* Firstly, all system variables have to be declared, that is the definition of each variable's type, e.g. binary, integer, real etc. Additionally, an extra integer variable Q denoting system's state is necessary.
- *Initial conditions.* After declaration, all variables have to be initialized in accordance to automaton's initial conditions. Additionally, variable Q takes the proper value denoting the initial state q_0.
- *Main Program.* The first two steps comprise the groundwork for the whole project, but building the executable program is the most important step. This step is separated in three stages where automaton's main actions take place.
- *Active state selection.* The whole program contains script reproducing all automaton's states and hence there must be a mechanism for choosing the active one and executing only the relative script.
- *Commands representing state's flow conditions.* In each state there must be all the flow conditions as given in the automaton. Of course, because in general an automaton includes system's model and the control strategy, only flow conditions representing system's controller are needed.
- *Commands representing state's transitions.* All the active events in each state have to be included in state's script. In general, a transition is triggered by an event, under the condition that is not ignored, and causes the change of the active state and the reset of some variables in accordance to its reset table.

The IEC 61131-3 standard has a set of five languages for programming PLCs which are supported by several manufacturers. The most common languages are the Instruction List (IL), the Ladder Diagram (LAD) and the Function Block Diagram (FBD) and they are generally supported by almost all manufacturing firms. Additionally, there are the Structured Text and the Sequential Function Charts (SFC), which are high level languages but they are not supported by all PLC types. The synthesis tool overview is depicted in Figure 5.7, where the above mentioned programming steps are given in accordance to each language's semantics. Accomplishing with tool's presentation, a more detailed description is given below.

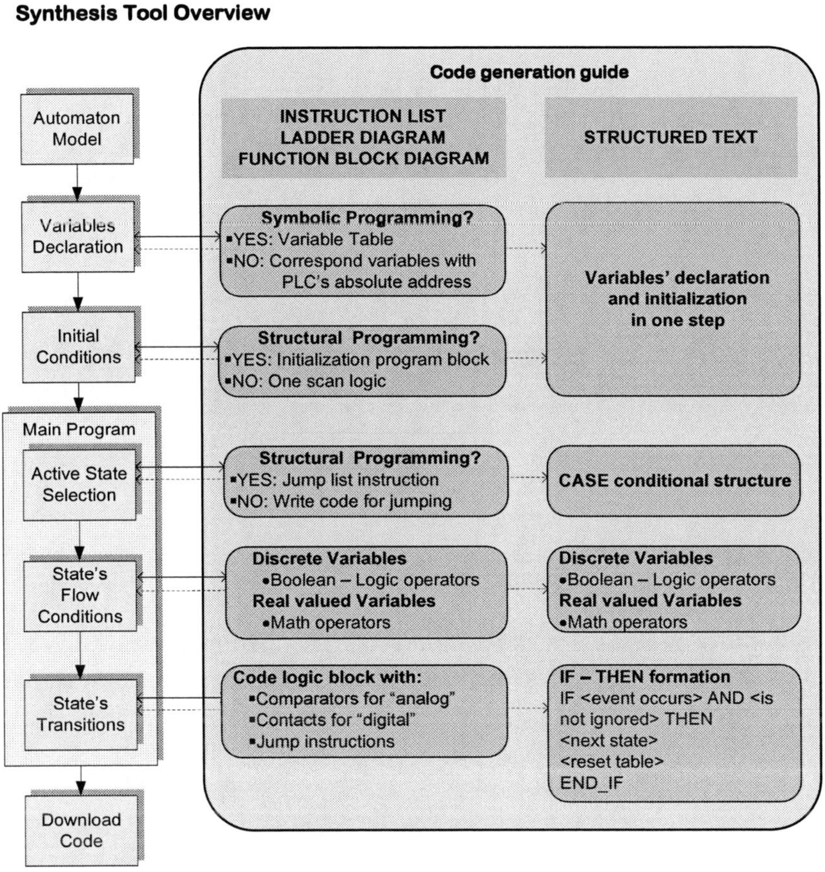

Figure 5.7. Synthesis tool overview.

5.2.1. Guide For Programming PLCs In IL, LAD and FBD

The guide for these three languages is given jointly, since all the commands used can be translated from one language to another. Many programming environments (e.g. STEP 7 for Siemens PLCs) which support these languages have special features for translating programs from one form to another. The directions are in accordance to the general steps as described in t he previous paragraph and are given based on Siemens SIMATIC STEP7 programming environment, in order to introduce specific functional examples.

- *Variables declaration.* All system variables are declared using the relative feature that the PLC's programming environment provides (e.g. Variable Table for STEP7).
- *Initial conditions.* All system variables are initialized by making use of the operation block (OB100) which STEP7 provides and is executed only once after system's restart. An example of the code required in IL is shown in Figure 5.8.

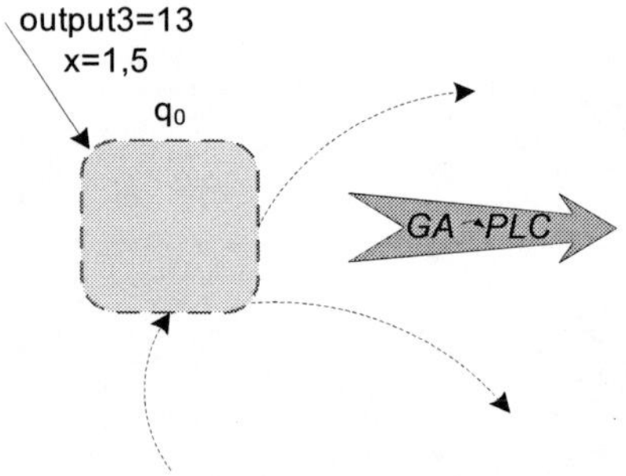


```
// OB100
L     0          // Set initial state q0
T     "q"
L     13         //Initialize integer variable output3=3
T     "output3"
L     1.5        //Initialize real variable x=1.5
T     "x"
END
```

Figure 5.8. Setting initial conditions in IL.

- *Main Program.* In order to have a structured program which will be optimal executed, a hierarchical structure is used. Particularly, the operation block OB1 is used for choosing the active state, which is executed as a Function Block (FB). Subsequently, one function block for each automaton's state is needed. The three parts of the main program are given below. The first one is implemented in the OB1, while the other two are implemented in each state's FB.
- *Active state selection.* In the operation block OB1 a Jump to List (JL) instruction is created, relatively to variable's q value. The JL instruction is a jump distributor and it is followed by a series of unconditional jumps to labels. Figure 5.9 shows the program and the relative flowchart. Each segment calls the corresponding FB which reconstructs state's operation.

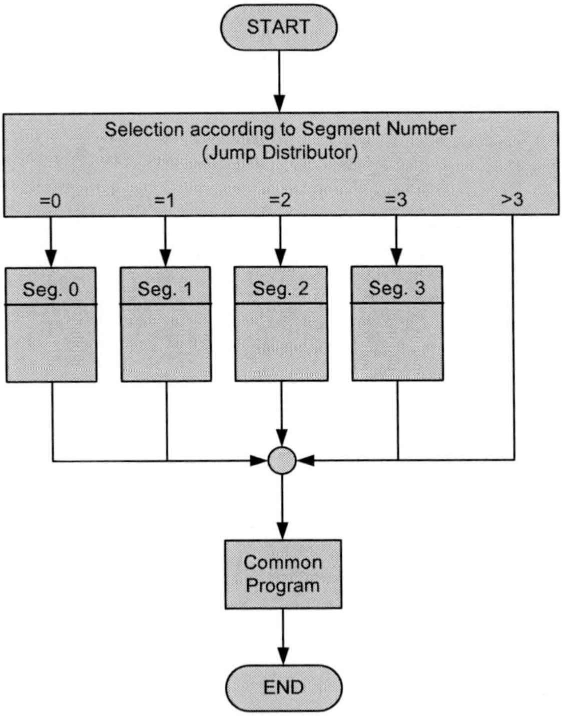


```
       ...
       L      "q"
       JL     LIST
       JU     SEG0
       JU     SEG1
       JU     SEG2
       JU     SEG3
LIST:  JU     COMM

SEG0:  CALL   "FB_State_0"
       JU     COMM

SEG1:  CALL   "FB_State_1"
       JU     COMM

SEG2:  CALL   "FB_State_2"
       JU     COMM

SEG3:  CALL   "FB_State_3"
       JU     COMM

COMM:  ...

       END
```

Figure 5.9. Selecting the active state using the Jump to List instruction.

- *Commands representing state's flow conditions.* As noted above, each state must include all the flow conditions as given in the automaton. Regarding discrete variables, flow conditions are Boolean equations and hence they can easily translated to code for each one of the used languages. On the contrary, flow conditions over the real valued variables are in general functions over time or differential equations. These equations can be transformed into difference equations [65], which are the analogues of differential equations in the finite difference calculus. In order to construct these equations we need the mathematical operators provided by the language (e.g. +R for adding real numbers, /I for dividing integers, SIN, EXP, LN, SQRT etc.). In case of FBD or LAD, mathematical equations are built with the embedded blocks. An example of calculating the result of x=sin(13*y)+k is given in Figure 5.10. Of course, any special features, such as PID or fuzzy controllers, can be used in order to simplify the overall task. Here, it has to be pinpointed that in each state's FB the commands representing state's flow conditions must be placed on the top and afterwards the commands representing state's transitions.

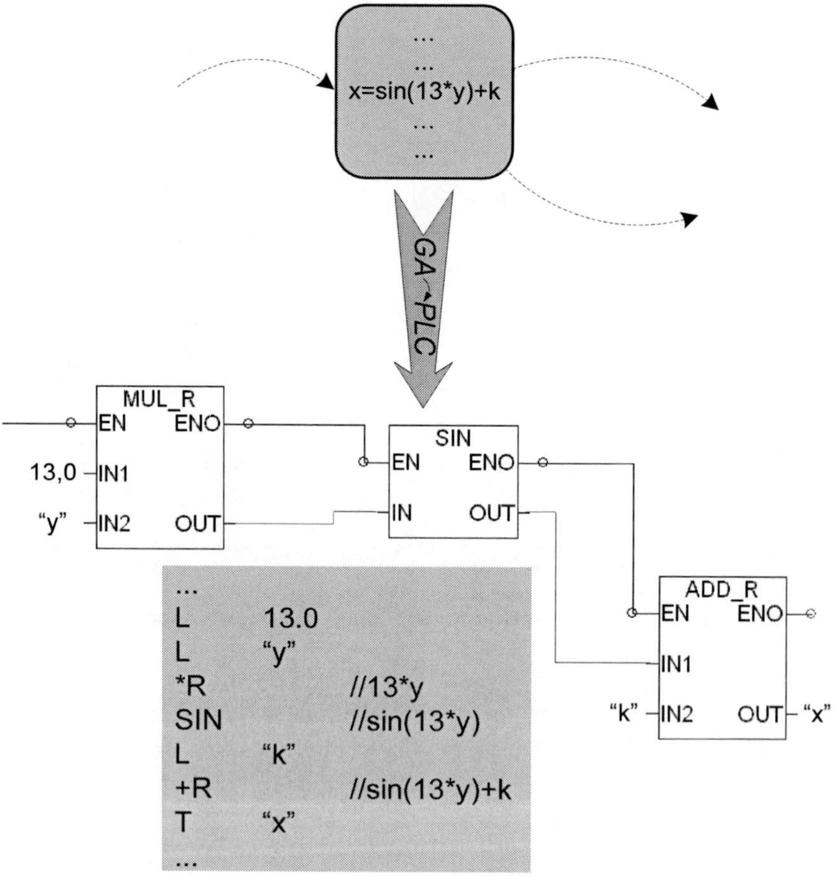

Figure 5.10. IL and LAD example.

- *Commands representing state's transitions.* All the active events in each state have to be included in state's script. Normally, a transition is triggered by an event (e.g. I1=TRUE), under the condition that is not ignored (e.g. x>23), and causes the change of the active state (e.g. q=3) and the reset of some variables in accordance to its reset table (e.g. k=2 and f=TRUE). The branches of reset table and active state change are MOVE structures. Moreover, in order to avoid the execution of another transition we need an unconditional jump to FB end. Event priority is denoted by the order in the script, that is, the event with the higher priority goes first while the event with the lower priority last. An example of a transition in LAD and FBD is given in Figure 5.11.

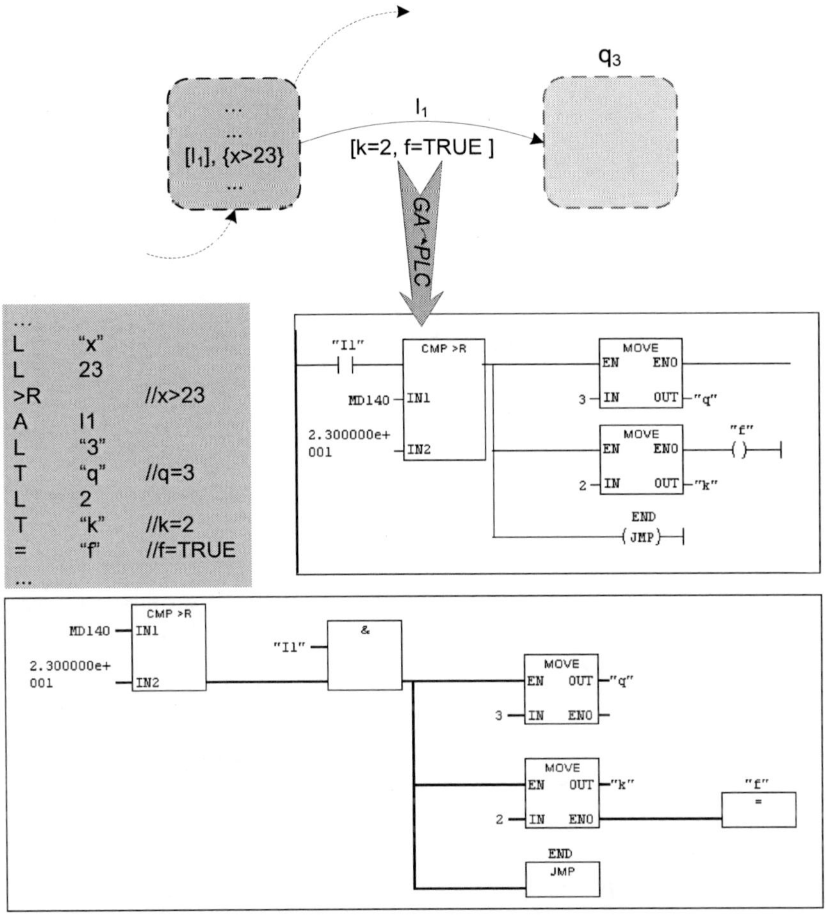

Figure 5.11. State transition in IL, LAD and FBD.

5.2.2. Guide for Programming PLCs in Structured Text

Structured Text (ST) is a high level textual language whose syntax is at first sight very similar to PASCAL. Although there are some similarities to PASCAL, ST is a distinct language that has been specifically developed for industrial control applications.

- *Variables declaration – Initial conditions.* When programming in ST variables declaration and initialization can be made jointly. Declaration takes place once again in OB100 while simple commands give the variable type and initial value (Figure 5.12).

```
e.g.   q : INT :=1;          (*Initial state q1*)
       X1 : REAL := 13;      (*Real valued variable X1=13*)
       X4 : BOOL := 0;       (*Binary variable X4=FALSE*)
```

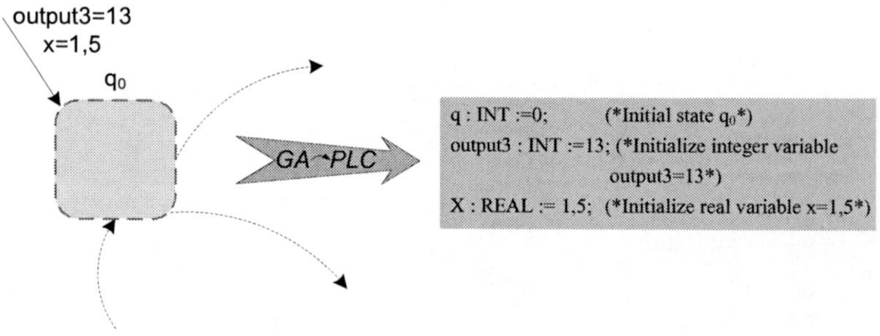

Figure 5.12. Variables declaration in ST.

- *Main Program.* In each execution loop the main program is executed in two blocks. The first one is the OB1 where the active state selection takes place, while the latter is the function block reconstructing the active state which includes its flow conditions and events. The commands representing state's flow conditions are given first and then are given the commands representing state's transitions.
- *Active state selection.* The mechanism for choosing the active state is a CASE conditional statement. In this structure, the set of statements which have an integer selector value that matches the value of the integer expression are executed. The integer expression used is the variable q which denotes the automaton's active state and the statement which calls the relative function block is executed.

```
e.g.    CASE  q     OF     (*Selection according to q value*)
              1 :   FB1;   (*If q=1, call FB1*)
              2 :   FB2;   (*If q=2, call FB2*)
              ...
              n :   FBn;   (*If q=n, call FBn*)
        END_CASE;
```

- *Commands representing state's flow conditions.* In each state there must be all the flow conditions as given in the automaton. Boolean equations are given by using ST logic operators (e.g. AND, &, OR, XOR, NOT etc.). Flow conditions over the real variables are built as difference equations with the relative math operators (Figure 5.13).

```
e.g.    Area := PI * R * R;
        Delta := Delta + 0.2*R;
        V := K ** (-W*T);
        Start := Valve AND Steam AND Pump;
        Status := (Valve1 = Open) XOR (Valve2 = Shut);
```

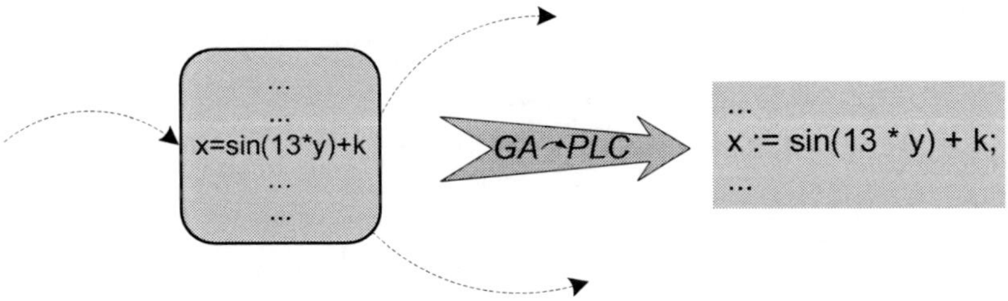

Figure 5.13. State's flow conditions.

- *Commands representing state's transitions.* Active events have to be included in state's function block, taking an IF – THEN formation. Each transition is caused by an event, if it is not ignored, and results in the state change and the reset of some variables. If there are more than one events in a state, relative IF – THEN structures are given in compliance with set Σ_i, which represents hierarchy of active events in i state. The general form of a transition construct is given below and an example with indicative data is given in Figure 5.14.

```
IF <event_occurs> AND <is_not_ignored> THEN
        <next_state>;
        <reset tables>;
END_IF


e.g.    IF (pump = ON) AND (temp > 200) THEN
                q := 4;
Valve1 := off;
                X2 := 100;
        END_IF
```

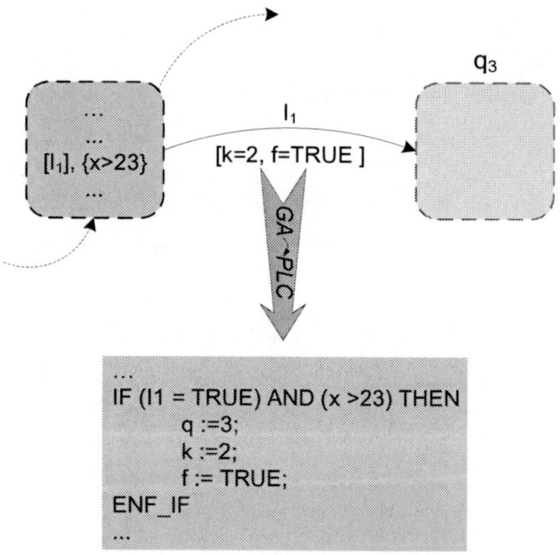

Figure 5.14. State's transitions.

5.2.3. Guide for Programming PLCs in Sequential Function Chart

Sequential Function Chart is a powerful graphical technique for describing the sequential behaviour of a control program. SFC can be used at the top level to show the main phases of a process with combination to other languages for the lower levels. Of course SFC can also be used at any other level, without the need of other languages. In this paragraph are given only two examples of implementing active state selection and transitions, because all the actions in SFC can be defined by using FBD, LAD or ST and examples of such actions are already given in previous paragraphs. Figure 5.15 gives the formation of the structure in the OB1 which selects the function block to be executed in accordance to variable's Q value (example with only four states).

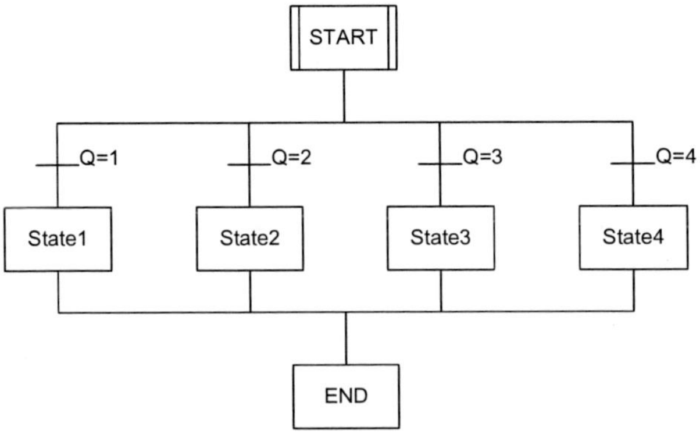

Figure 5.15. OB1 in SFC.

Respectively, Figure 5.16 presents the structure of a transition with the event, the relative restriction and the state change with the reset table.

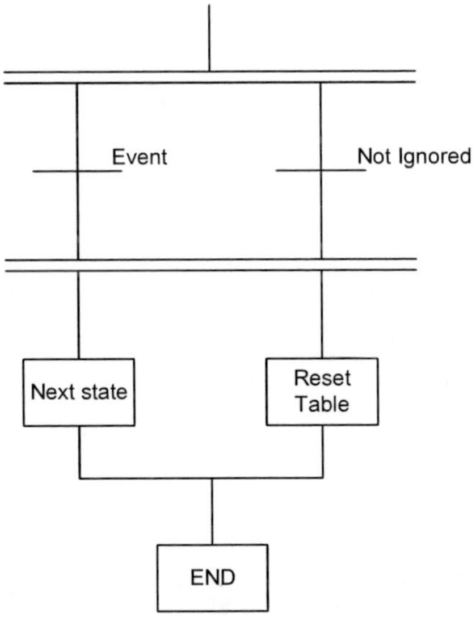

Figure 5.16. A transition structure in SFC.

5.2.4. In Case Of Multiple Automata Model.

In some cases, the overall plant is modelled by two or more automata and hence, the controlled system demands the parallel execution of them. Such an example is given in section 6.2, where the system is a concrete batching and mixing plant. The whole system is modelled by five different automata executed simultaneously. Compared to single automata systems, the only difference is the active state selection mechanism, which selects active states of each automaton. The selection hierarchy (i.e. hierarchy of calling state functions) represents execution hierarchy between automata.

5.3. Applications of Guide Use in Real Problems

In this section, guide's use is presented through an simple, but representative, example. The described system implements an electronic lock with a controller specified by the means of two binary variables *X* and *Y*. When the system operates, is set at an initial state, from where the evolution of *X* and *Y* is watched and determines system's operation. The lock must open if *X* and *Y* are operated in the following sequence:

1. First *X* is activated.
2. Then *Y* is activated and deactivated.
3. Here, the first output is activated, unlocking the first kingpin K_1.
4. Then *X* is deactivated.
5. Finally, *Y* is activated and deactivated unlocking and the second kingpin K_2 and door opens.

If *X* and *Y* are activated in the wrong sequence, the system returns to its initial state, where both kingpins' outputs are deactivated. It also returns to the initial state by the action of a micro-switch *M* indicating the time when the door was shut. Figure 5.17 shows the state diagram of the described system which consists of seven states. State q_0 is the initial state where the system is set when power is switched on. States q_1 to q_5 are intermediate states before the system reaches state q_6 where door opens.

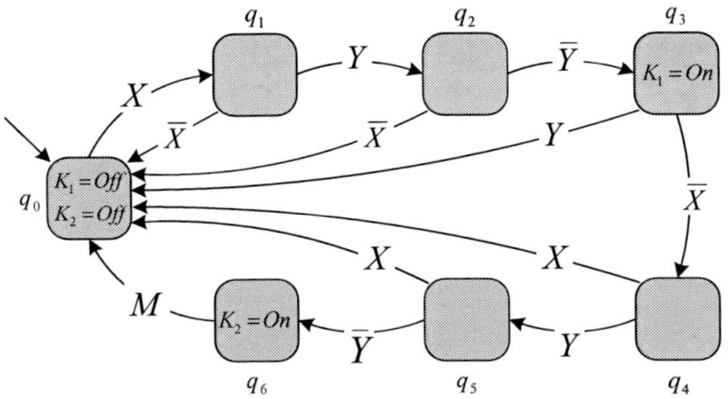

Figure 5.17. Smart lock Global Automaton.

An implementation in a PLC using the Instruction List programming tool is shown in Figure 5.18. A jump to list structure was used and the whole program is simple enough since the system is discrete without any continuous variables which in general increase system's complexity.

```
      L    "q"
      JL   LIST
      JU   ST0
      JU   ST1
      JU   ST2
      JU   ST3
      JU   ST4
      JU   ST5
      JU   ST6
LIST: JU   COM
ST0:  R    "K1"
      R    "K2"
      A    "X"
      L    "1"
      T    "q"
      JU   COM
ST1:  AN   "X"
      L    "0"
      T    "q"
      JC   COM
      A    "Y"
      L    "2"
      T    "q"
      JU   COM

ST2:  AN   "X"
      L    "0"
      T    "q"
      JC   COM
      AN   "Y"
      L    "3"
      T    "q"
      JU   COM
ST3:  S    "K1"
      A    "Y"
      L    "0"
      T    "q"
      JC   COM
      AN   "X"
      L    "4"
      T    "q"
      JU   COM
ST4:  A    "X"
      L    "0"
      T    "q"
      JC   COM
      A    "Y"
      L    "5"
      T    "q"
      JU   COM

ST5:  A    "X"
      L    "0"
      T    "q"
      JC   COM
      AN   "Y"
      L    "6"
      T    "q"
      JU   COM
ST6:  S    "K2"
      A    "M"
      L    "0"
      T    "q"
      JC   COM
COM:  END
```

Figure 5.18. Smart lock implementation in IL.

6. EXAMPLES OF USING GLOBAL AUTOMATA

In order to thoroughly present the use of Global Automata, three detailed examples are presented in this section. The first one gives the model of a reciprocating internal combustion engine, a system largely studied in the hybrid automata bibliography. Following, a complete concrete batching and mixing plant is introduced, as a small-scale real industrial system. Finally, the third example is a three tank system, which has been implemented in a laboratory of University of Patras.

6.1. Reciprocating Internal Combustion Engine

The reciprocating internal combustion engine must be by far the most common form of engine or prime mover. As with most engines, the usual aim is to achieve a high work output with a high efficiency. The two main types of internal combustion engines are: Spark Ignition (SI) engines, where the fuel is ignited by a spark; and Compression Ignition (CI) engines, where the rise of temperature and pressure during compression is sufficient to cause spontaneous ignition of the fuel. The SI engine is also referred to as the petrol, gasoline or gas engine from its typical fuels, and the Otto engine, after the inventor. The CI engine is also referred to as the Diesel or oil engine.

During each crankshaft revolution there are two strokes of the piston, and both types of engine can be designed to operate in either four strokes or two strokes of the piston. In this paper, we are dealing only with the four-stroke spark ignited engine. The four-stroke operating cycle [66, 67, 68, 69] can be explained by reference to Figure 6.1, which shows a SI engine's cylinder inside.

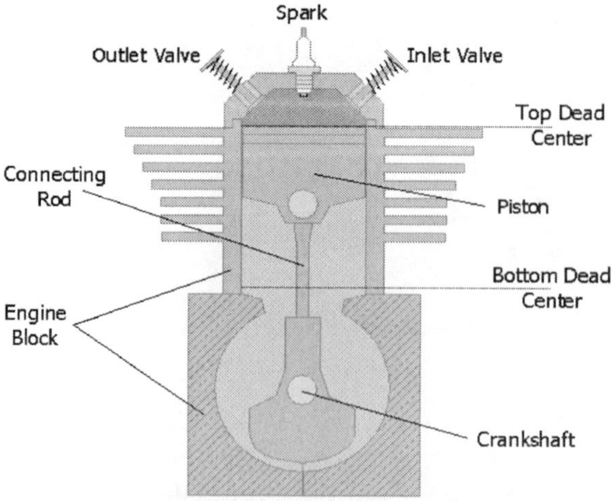

Figure 6.1. Cylinder of a SI engine.

1) *Induction or Suction Stroke:* The inlet valve is open, and the piston travels down the cylinder, drawing in a charge of air. In the case of a SI engine the fuel is usually pre-mixed with the air.
2) *Compression Stroke:* Both valves are closed, and the piston travels up the cylinder. As the piston approaches top dead center (TDC), ignition occurs. In the case of CI engines, the fuel is injected towards the end of the compression stroke.
3) *Expansion, Power or Working Stroke:* Combustion propagates throughout the charge, raising the pressure and temperature, and pushes the piston down forcing the crank to rotate. About five times as much work is done on the piston during the power stroke as the piston had to do during compression. At the end of the power stroke the exhaust valve opens, and the irreversible expansion of the exhaust gases is termed 'blow-down'.
4) *Exhaust Stroke:* The outlet valve remains open, and as the piston travels up the cylinder the remaining gases are expelled. At the end of the exhaust stroke, when the outlet valve closes some exhaust gas residuals will be left; these will dilute the next charge.

Whether an internal combustion engine operates on a two-stroke or four-stroke cycle and whether it uses spark ignition or compression ignition, it follows a mechanical cycle not a thermodynamic cycle. However, the thermal efficiency of such an engine is assessed by comparison with the thermal efficiency of air standard cycles, because of the similarity between the engine indicator diagram and the state diagram of the corresponding hypothetical cycle. The engine indicator diagram is the record of pressure against cylinder volume, recorded for an actual engine. Pressure/volume diagrams are very useful [11], as the enclosed area equates to the work in the cycle.

The Otto cycle [9] is usually used as a basis of comparison for SI engines. The cycle consists of four non-flow processes, as shown in Fig. 6.2. The compression and expansion processes are assumed to be adiabatic (no heat transfer) and reversible, and thus isentropic.

The physical description of an SI engine was given in section 2, and its air cycles were especially mentioned. In order to have a form convenient for analysis, a set of assumptions which simplify these two air-cycles has to be made.

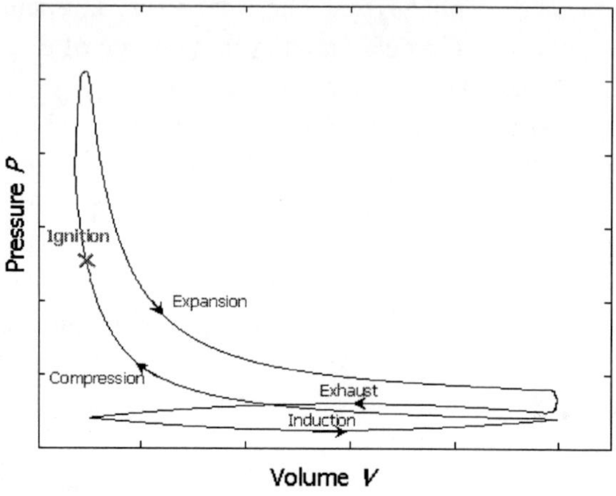

Figure 6.2. *p-V* diagram for a real four stroke engine.

The mathematical model for each stroke, derived from these assumptions, is:

1) *Induction or Suction Stroke:* Most models imply that induction takes place at constant pressure, but a more accurate model [70] would imply that:

$$p = f\left(V^2\right)$$

2) *Compression Stroke:* During compression the relation between pressure and volume, as shown in [71], can be described by:

$$p \cdot V^{1.3} = const.$$

3) *Expansion, Power or Working Stroke:* Combustion is assumed to be instantaneously and implies on pressure's abrupt rise. Expansion, as shown in [71], takes place under condition:

$$p \cdot V^{1.48} = const.$$

4) *Exhaust Stroke:* Exhaust stroke starts with outlet valve opening, which results in an abrupt pressure blow-down. After that, similarly to induction stroke, exhaust stroke usually is assumed to be at constant pressure, but more accurately [68, 70]:

$$p = f\left(-V^2\right)$$

Model's graphical representation is shown in Fig. 6.3.

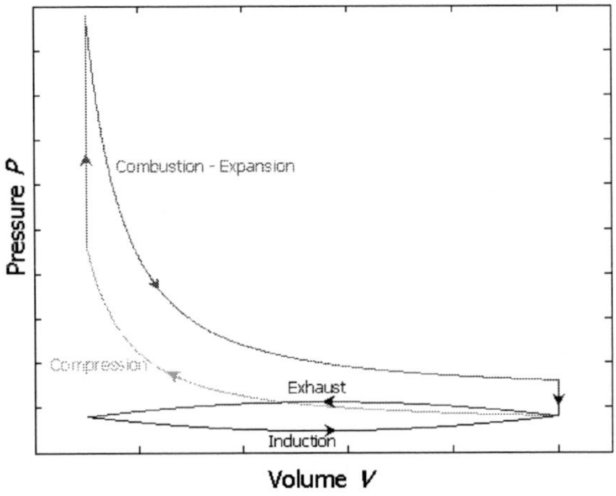

Figure 6.3. *p-V* diagram of mathematical model.

Regarding global automata, a SI engine can be modeled as an automaton with four states, one for each stroke. The approach used in simulation is shown in Figure 6.4. Each state has a relative set of parameters, which indicate valves' condition, spark's ignition and pressure's and volume's fluctuation. Transitions between states are triggered either when piston reaches Top Dead Center (TDC, volume is minimum) or when piston gets to Bottom Dead Center (BDC, volume is maximum). Transitions between Compression-Power and Power-Exhaust states have a reset table in order to model pressure's abrupt rise and blow-down.

The model described in previous section was used for simulation in CX – Supervisor, a SCADA software by OMRON. Two screens were designed, one for the animated representation of piston's move and automaton's run, and a second with graphical depiction of all system's variables. Automaton's parameters (states, transitions, flow conditions, reset table, etc.), were reproduced by an interval script written in CX – Supervisor's Editor, using the presented guide. All parameters are given as functions of crankshaft's angle, see Figure 6.5.

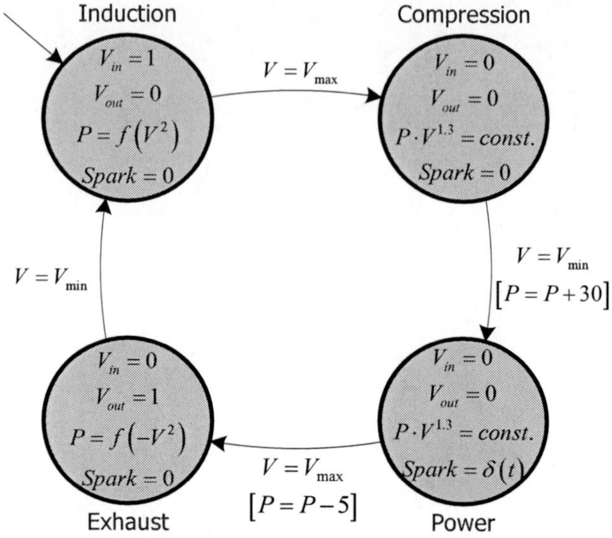

Figure 6.4. Engine's automaton with four states.

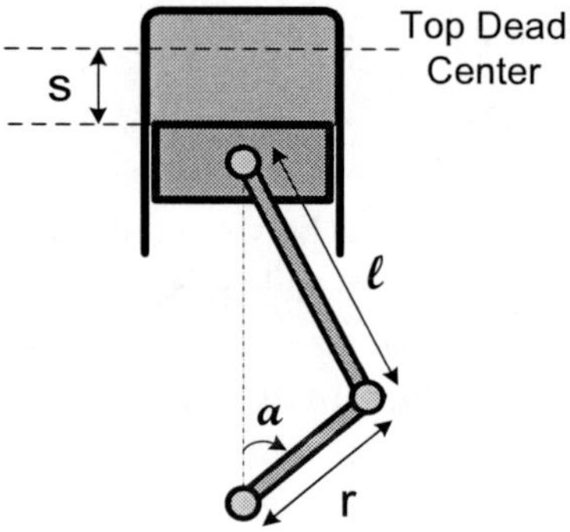

Figure 6.5. Crankshaft.

The piston displacement from TDC, which is analogous to cylinder volume, is:

$s(a) = r\left(1 - \cos\alpha + \frac{l}{r}\left(1 - \sqrt{1 - \frac{r^2}{l^2} sin^2 a}\right)\right)$, where a is the crankshaft's angle, r is crankshaft's radius and l is connecting rod length. Cylinder volume depends on engine's constructing parameters, such as bore, which is cylinder diameter, and cylinder length. At TDC, we have $a = 0$, $s(a) = 0$ and $V = min$, whereas at BDC, $a = \pi$, $s(a) = 2 \cdot r$ and $V = max$ respectively.

The executable code for the Compression stroke, as derived using the guide, is given here below:

```
CASE 2
   Vin=0
   Vout=0
   Spark=0
   V=K*r(1-cos(a)+(l/r)*(1-sqrt(1-r^2*(sin(a))^2/l^2))
   P=K2/V
   a=a+pi/180
          if V==Vmin then
          q=3
          P=P+30
          endif
```

The two above mentioned screens are shown in Figures 6.6 and 6.7.

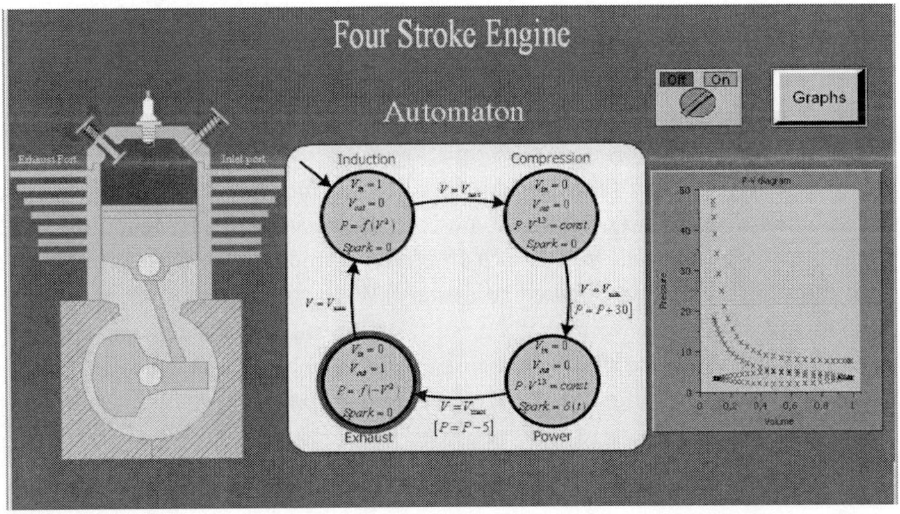

Figure 6.6. Screenshot of engine's simulation.

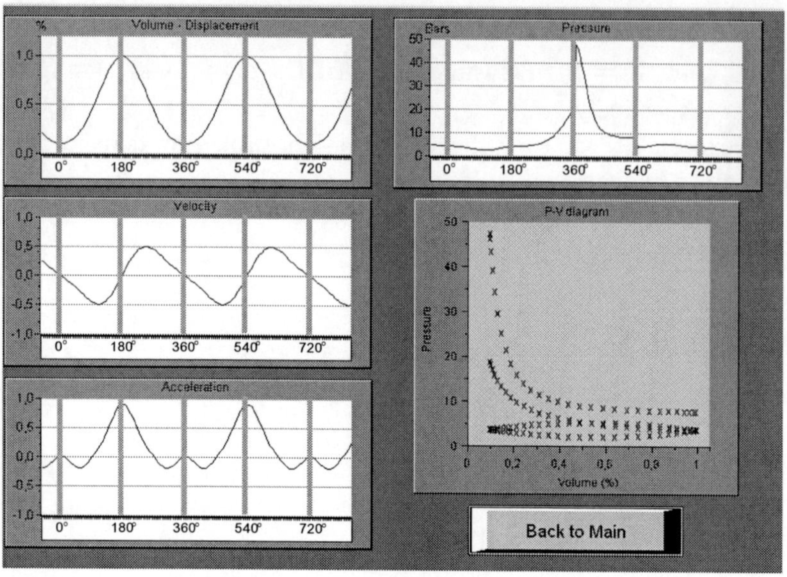

Figure 6.7. Graph-page screenshot.

6.2. Concrete Batching and Mixing Plant

The central mix plant studied [72] in this project is depicted in Fig. 6.8. Batching is the process of weighing and introducing into the mixer the ingredients for a batch of concrete. All raw materials, cement, aggregates, water and admixtures are weighted and batched, at the aggregate and cement weigh hoppers and at the admixtures reactor respectively. Our plant has five different types of aggregates each one stored in separate feeder bin. Cement composition types are three, placed in silos. Weigh hoppers have only one weigh meter and hence each type of aggregates or cement is added consequently and not all together. Admixtures, which are in liquid form, have also three different types tanked separately. Water is controlled using a meter system measuring the volume required. For the mixing process a twin shaft mixer is used to ensure that concrete is mixed thoroughly until it is uniform in appearance and all ingredients are evenly distributed. Finally, the freshly mixed concrete is deposited into trucks for delivery to the site of work.

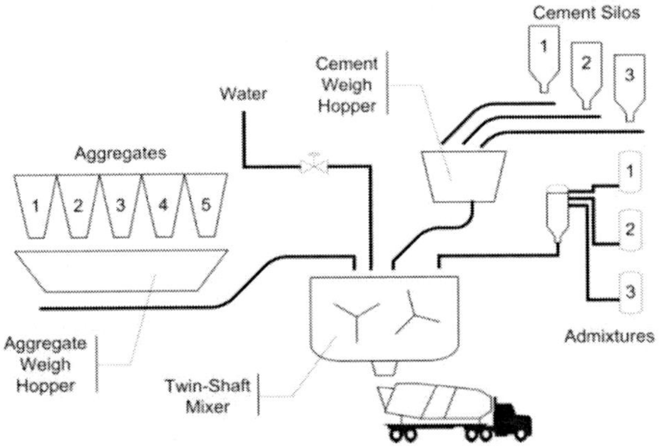

Figure 6.8. Schematic of concrete batching and mixing plant.

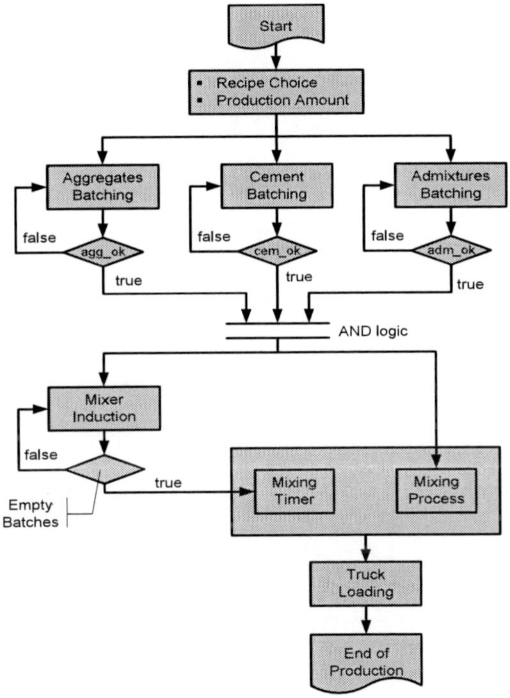

Figure 6.9. Batching and mixing process flowchart.

The most crucial factor concerning concrete type and its mechanical and chemical properties is the materials proportion. Furthermore, other essential parameters such as material hierarchy - regarding the input sequence in the mixer - and the mixing time need to be considered in order to ensure the quality objective. All parameters regulated or not, are specified in the corresponding recipe specification sheet used for each concrete mix.

A flowchart of the overall process is depicted in Fig. 6.9, where all the actions for batching and mixing are shown. When production starts, system operator has to choose the desired recipe and insert the wanted quantity of concrete through the SCADA graphic user interface. Then, the batching process starts for the three different kinds of materials (there is no need for batching water), and process continues to the next step, which is the mixing process, only when all ingredients are batched properly. Mixing process starts simultaneously with the material induction in the mixer. Induction hierarchy, as noted above, depends on the specific recipe chosen. Mixing timer starts counting mixing time after all ingredients are placed into the mixer and when time elapses, mixing process stops and truck is loaded with concrete. When truck loading ends and mixer is empty, the production is completed and the system is ready for a new one.

6.2.1. Modeling and Control Using Global Automata

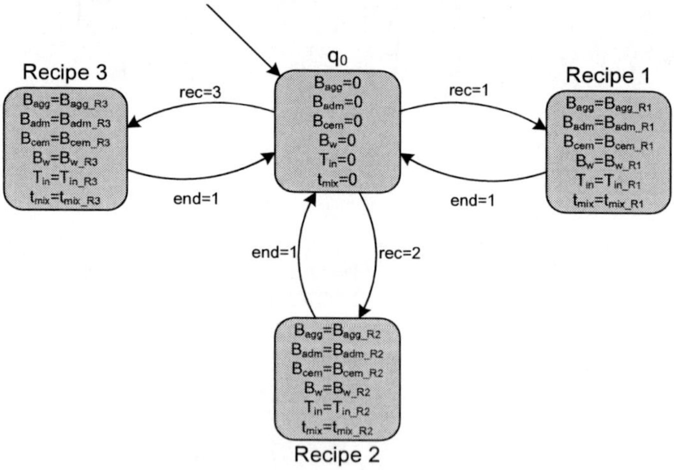

Figure 6.10. Supervisor - Recipe automaton.

There are several approaches in bibliography for modeling similar system [73, 74, 75]. The approach used in this project implies the use of several automata each one modeling a specific part of the overall system. First of all, an automaton describing the recipe mechanism has to be constructed. This automaton plays the role of the supervisor controller feeding all the other controllers with specific set-points in order to achieve the desired proportion of raw materials. The production's supervisor is depicted in Fig. 6.10. State q_0 is the initial state, while all the other states represent a relative recipe; in order to have a legible figure, only three recipes are given. Each recipe gives the variable values which are essential for the whole procedure. Firstly, the proportions of each material have to be declared, $B_{agg} = \{b_{agg\,1}, b_{agg\,2}, b_{agg\,3}, b_{agg\,4}, b_{agg\,5}\}$, $B_{cem} = \{b_{cem\,1}, b_{cem\,2}, b_{cem\,3}\}$, $B_w = b_{water}$ and $B_{adm} = \{b_{adm\,1}, b_{adm\,2}, b_{adm\,3}\}$. Furthermore, material hierarchy relatively to the import sequence in the mixer is given, $T_{in} = \{t_{agg}, t_{adm}, t_{cem}, t_w\}$. Finally, mixing time is also determined in the supervisor automaton, t_{mix}.

Thereafter, one automaton for the batching process of each material category is needed. Hence, three automata were developed for modeling aggregate and cement weigh hoppers as well as admixtures reactor. These automata are shown in Fig. 6.11, 6.12 and 6.13 respectively.

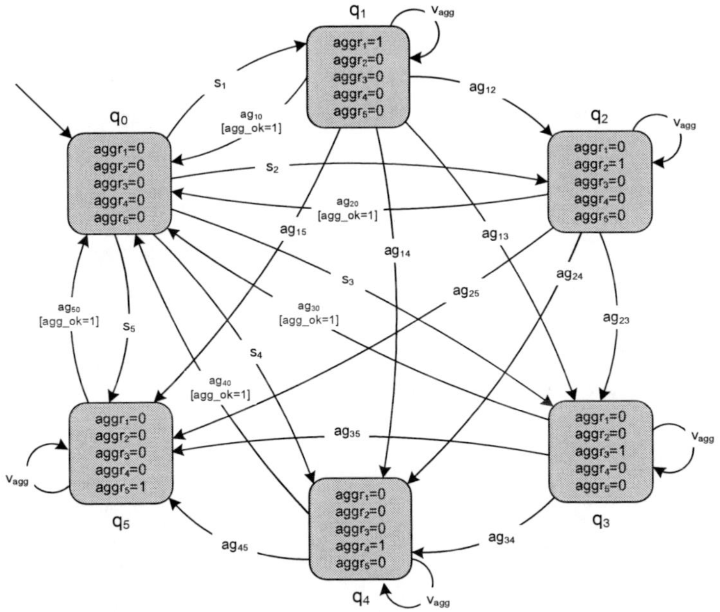

Figure 6.11. Automaton for aggregate weigh hopper.

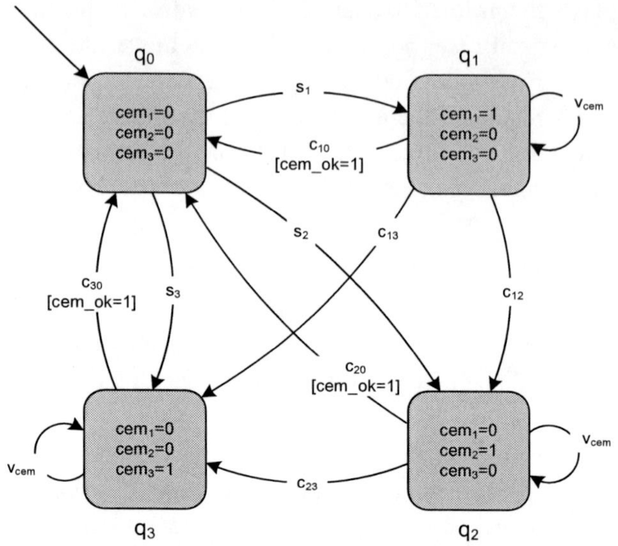

Figure 6.12. Automaton for cement weigh hopper.

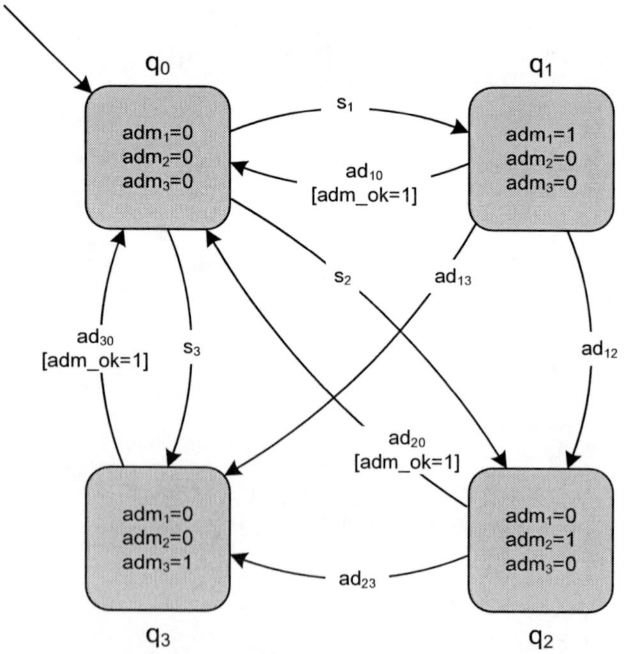

Figure 6.13. Automaton modelling admixtures reactor.

The three batching automata are similar since their main operation is to weigh out the right proportions for each different material. The notation at each figure has to be clarified, but due to their similarity only the notation of Fig. 6.11 is given. Firstly, q_0 is the initial state where all the feeder bins' doors are closed (variables $aggr_i$). Each state q_i, i=1,…,5 implies that the doors of the ith feed bin are open. Transitions from the initial state labeled with s_i are triggered by the event $start \cdot \left(b_{agg\,1} > 0\right)$ and cause the state transition to the q_i state. Respectively, transitions ag_{ij}, i=1,…,4 and j=2,…,5 are triggered by the events $\left(B_{agg\_tot} = b_{agg\,1} + \cdots + b_{agg\,i}\right) \cdot \left(b_{agg\,i+1} = 0, \ldots, b_{agg\,j-1} = 0\right) \cdot \left(b_{agg\,j} > 0\right)$ and cause the state transitions between the state q_i and q_j. In example, transition ag_{25} is triggered when event $\left(B_{agg\_tot} = b_{agg\,1} + b_{agg\,2}\right) \cdot \left(b_{agg\,3} = 0, b_{agg\,4} = 0\right) \cdot \left(b_{agg\,5} > 0\right)$ occurs. This transition is executed only if the recipe chosen implies that the amounts of aggregates 3 and 4 are null. Transitions marked as ag_{i0} are caused by the events $\left(B_{agg\_tot} = b_{agg\,1} + \cdots + b_{agg\,i}\right) \cdot \left(b_{agg\,i+1} = 0, \ldots, b_{agg\,j-1} = 0\right)$. These transitions have also a reset table [*agg_ok=1*] as a confirmation signal that all the aggregates are weighed and batched in the weigh hopper properly. Finally, regarding events hierarchy in each state it is as follows: (q_0: s_1, s_2, s_3, s_4, s_5), (q_1: ag_{12}, ag_{13}, ag_{14}, ag_{15}, ag_{10}, v_{agg}), (q_2: ag_{23}, ag_{24}, ag_{25}, ag_{20}, v_{agg}), (q_3: ag_{34}, ag_{35}, ag_{30}, v_{agg}), (q_4: ag_{45}, ag_{40}, v_{agg}) and (q_5: ag_{50}, v_{agg}). Furthermore, each material flow rate is calculated and if it is low, a relative vibrator turns on for a few seconds (loop transitions labeled v_{agg}). Automata for cement and admixtures are alike weighing and batching the specified amounts of each ingredient.

Finally, an automaton modeling mixer's operation was needed. This automaton is depicted in Fig. 6.14 and its main operation is to feed mixer with all the ingredients and control the mixing process. State q_0 is the initial state and after all ingredients are weighed and batched properly (signal $agg\_ok \cdot adm\_ok \cdot cem\_ok$) a state transition to q_1 takes place. Here, it must be noted that in the given approach, mixing timer starts simultaneously with the mixing process, but as shown in the flowchart of Fig. 6.9, timer starts when all components are discharged into the mixer. In this case, we have a slightly diversified approach where mixing time includes the time needed for mixer feeding.

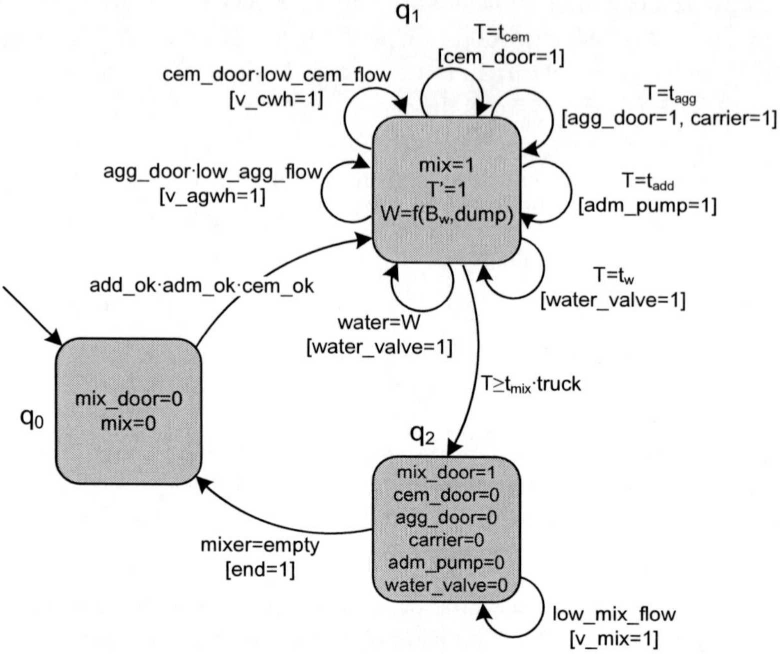

Figure 6.14. Automaton modelling mixer's operation.

6.2.2. Supervisory Control and Data Acquisition Station

The model and the control design were verified using a medium-scale real plant, the batching and mixing process of which incurs the risk of significant loss in case of an error or system failure. The whole system's controller was implemented in CX–Supervisor, a SCADA software by OMRON Corp. Taking advantage of software's animation capabilities, screens resembling real system were constructed in order to visualize system's operation. The main screen for the overview of the whole system is given in Fig. 6.15. Figure 6.15 presents the graphical representation of the concrete plant with all its subsystems. All the batching subsystems are clearly presented and the animation features show the procedure's evolution. Animation includes carriers, motors, mixer's operation and materials flow.

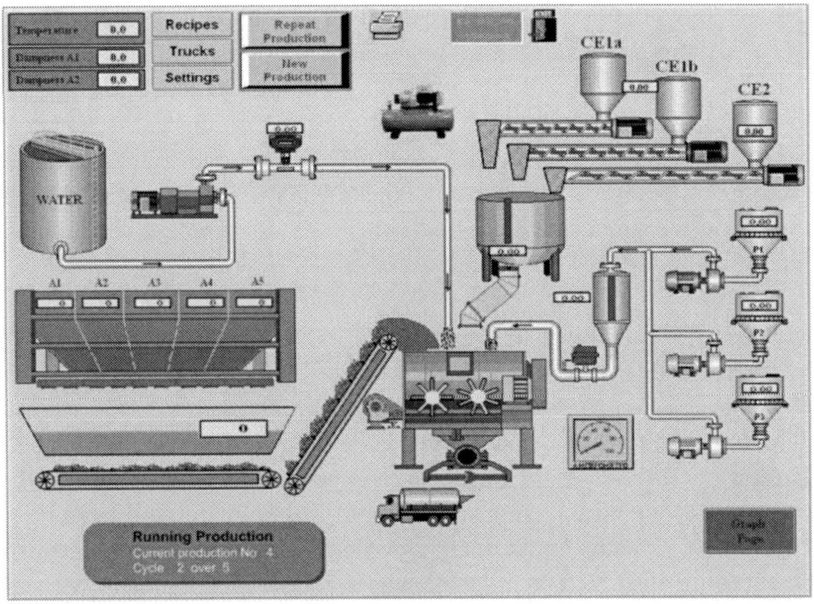

Figure 6.15. SCADA homepage of the batching and mixing plant.

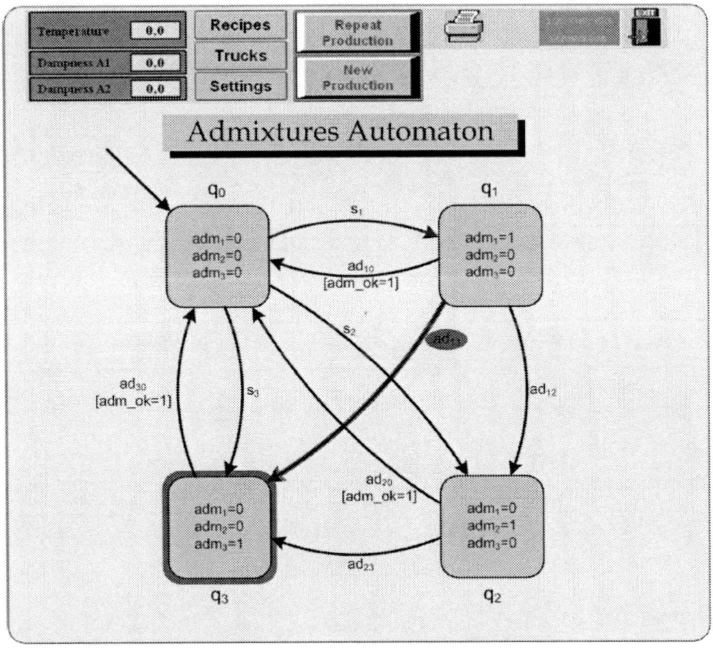

Figure 6.16. Admixtures automaton's run simulation.

Except from the main SCADA page with the graphical reconstruction of the system, there are several other pages showing the automata evolution. An example is given in Fig. 6.16, where the automaton modeling admixtures reactor is shown. The active state (q_3) on present time instant is noted with a red square, as also the last event (ad_{13}) and the relative triggered transition. The automaton being in state q_3 means that the admixture No. 3 is added to the reactor while the precedent state was q_1 since the recipe chosen implies null amount of admixture No. 2 to the mixture.

6.3. A Three Tank System

The final example of using the proposed method has been implemented in a laboratory of University of Patras. The experimental setup consists of three independent tanks and the control issue is to fill the main tank from the other two according to some predefined criteria. The two tanks contain different liquids feeding the third main tank where the mixing process takes place. System's controller is a Siemens S7-300 PLC, which gathers information from the plant through the relative level sensors and implies the control policy to the pumps and the analogue valves assuring the control criteria are satisfied. A plant schematic is shown in Figure 6.17 while in Figure 6.18 a photo of the whole setup (tanks, valves, sensors, PLC) is given.

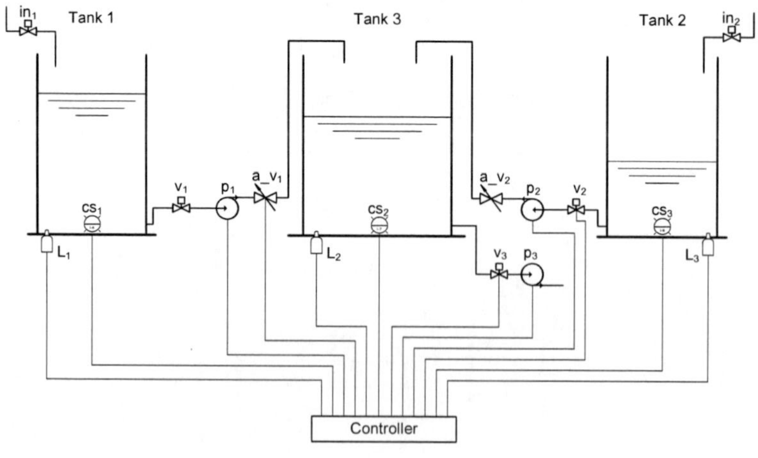

Figure 6.17. Plant's schematic representation.

Figure 6.18. Photo of the experimental setup.

The experimental setup resembles a paint factory where chromatic different liquids are mixed in order to produce a predefined colour shade. Each liquid's colour is measured through a colour sensor (cs_1, cs_2, cs_3) in the bottom of each tank. The whole procedure starts through a *Start* button. When *Start* button is pressed, the on/off valves v_1 and v_2 open, the pumps p_1 and p_2 start and the analogue valves a_v_1 and a_v_2 open in accordance to control strategy and both liquids flow to Tank 3. Simultaneously the mixing process starts by turning on the mixer, *mix*. When liquid mass in Tank 3 reaches 70% all valves and pumps close and depending from mixture's colour, the system adds liquid only from Tank 1 or Tank 2, or if the colour is acceptable the mixing process ends, the mixer *mix* turns off and Tank 3 empties by opening valve v_3 and turning on pump p_3.

The three tank system can be seen as a whole, or in a different approach, as three independent subsystems. Each subsystem models one of the three tanks using global automata. These three automata designed for each subsystem are shown in Figures 6.19, 6.20 and 6.21.

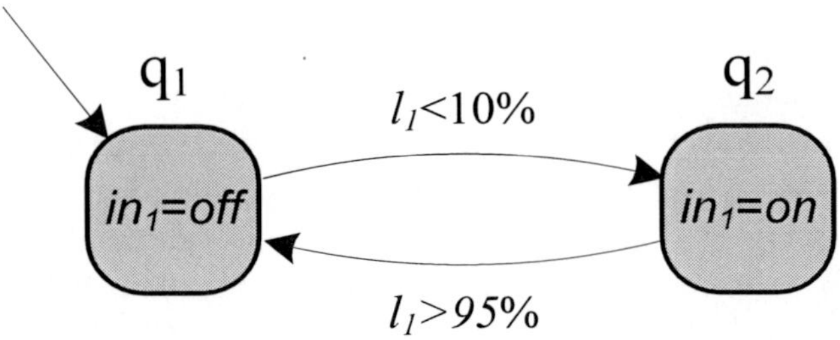

Figure 6.19. First automaton modelling Tank 1.

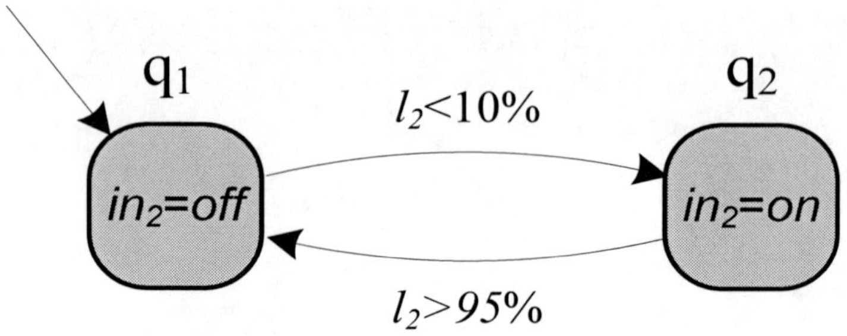

Figure 6.20. Second automaton modelling Tank 2.

The first two automata are very simple controlling tanks' inlet in_1 and in_2 in accordance to liquids level. The third automaton is quite complex since it encloses all the system's control strategy. It consists of six process states, one for each process phase. In additional there are two alarm states *ALARM* and *STOP*, where system goes, if the relative buttons are pressed. Regarding *STOP* state, it stops the process and closes every pump and valve, and if the *start* button is pressed, it goes back to the former state before the *stop* signal. *ALARM* state stops the process, empties Tank 3 and returns to initial state Q_1. Only for presentation reasons, transitions to *ALARM* state are given with dotted lines, while transitions to and from *STOP* state are given with dashed lines.

The process procedure is presented more analytically below. Initially the system is in state Q_1 and when button start is pressed a transition to state Q_2 takes place. In this state, both liquids flow to Tank 3 and the mixing process starts, until it reaches the upper limit of tank's containment, where the system goes to state Q_3. This is an intermediate state to measure the accuracy of mixture controller. If the mixture is in accordance to specifications, the transition to state Q_4 occurs. But, if the criteria are not fulfilled, the system transits to state Q_{31} or Q_{32}, where addition of liquid 1 or 2 takes place respectively, and then the system goes to final state Q_4, where the mixing process ends and the produced liquid flows from Tank 3 to the exit. When Tank 3 is completely empty, process stops returning to the initial state.

The program which implements the control strategy described by the three automata was written in STL language and a sample part, regarding sate Q_{31}, is given in Figure 6.22. The controller used in the specific state was a PI controller implemented using the relative embedded PLC function

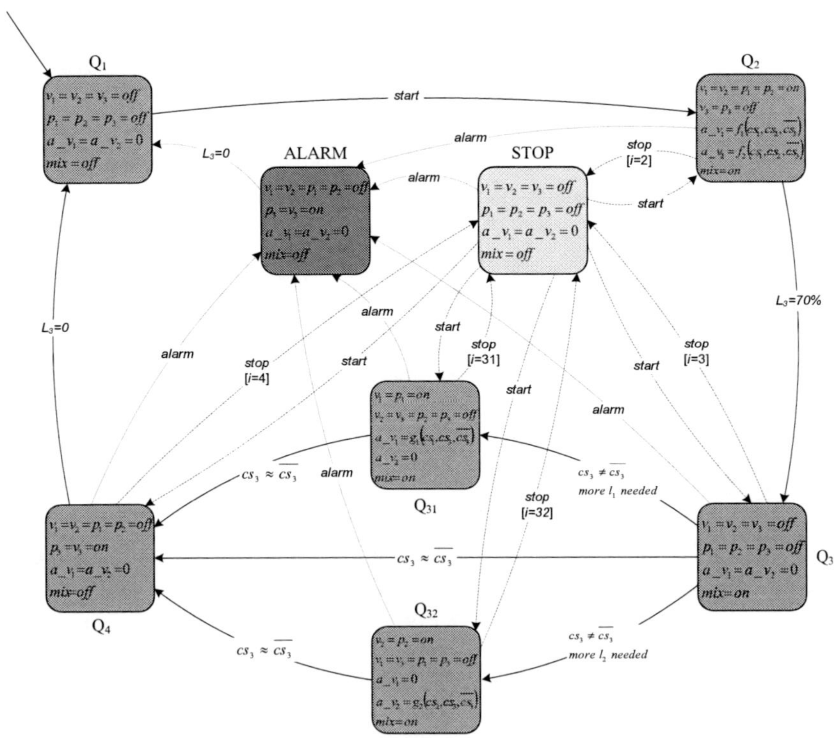

Figure 6.21. Third automaton modelling Tank 3(Green states are the regular process states, while yellow and red are the stop and alarm states respectively).

```
// Function Q31

  S      "V2"
  S      "P2"
  S      "MIX"
  R      "V1"
  R      "V3"
  R      "P1"
  R      "P3"
  L      0.0
  T      "A_V1"

CALL "CONT_C" , DB42
 COM_RST  :=FALSE
 MAN_ON   :=M123.1
 PVPER_ON:=TRUE
 P_SEL    :=TRUE
 I_SEL    :=TRUE
 INT_HOLD :=TRUE
 I_ITL_ON :=FALSE
 D_SEL    :=FALSE
 CYCLE    :=T#10MS
 SP_INT   :=MD135
 PV_IN    :=MD141
```

```
PV_PER    :=MW220
MAN       :=MD240
GAIN      :=MD260
TI        :=T#20S
TD        :=T#10S
TM_LAG    :=T#2S
DEADB_W   :=0.000000e+000
LMN_HLM   :=1.000000e+002
LMN_LLM   :=0.000000e+000
PV_FAC    :=1.000000e+000
PV_OFF    :=0.000000e+000
LMN_FAC   :=1.000000e+000
LMN_OFF   :=0.000000e+000
I_ITLVAL  :=0.000000e+000
DISV      :=0.000000e+000
LMN       :=MD300
LMN_PER   :=MW320
QLMN_HLM:=M121.1
QLMN_LLM:=M121.2
LMN_P     :=MD324
LMN_I     :=MD328
LMN_D     :=MD332
PV        :=MD336
ER        :=MD340
```

```
L      MD400 //cs3
L      MD440 //cs3_opt*0.95
>=R
=      M380.1

L      MD400 //cs3
L      MD460 //cs3_opt*1.05
<=R
=      M380.2

A      M380.1
A      M380.2
L      "4"
T      "Q"

A      "STOP"
L      "5"
T      "Q"

A      "ALARM"
L      "6"
T      "Q"

END
```

Figure 6.22. STL program regarding state Q_{31}.

Conclusions

Industrial systems are usually described as DESs or as hybrid systems and they are mostly controlled by Programmable Logic Controllers. Many modelling methods have been proposed and a recently proposed one is Global Automata. This paper gives a synthesis tool for implementing global automata in PLCs based on the IEC 61131-3 programming norm. The presented guide may become a helpful tool for industrial engineers who want to program a PLC in accordance to system's global automaton model. It describes every language of the IEC 61131-3 norm giving the engineer the ability to decide and choose the best one for each project or system. Because different PLC manufacturers use different syntax rules and have special attributes, the presented tool has more general directives on building such models.

The next step is the development of a new software tool for modelling, verification and simulation of industrial systems based on global automata. This tool must have a compiler, based on the presented guide, which will produce the executable program for several target PLCs, so as to eliminate engineer tasks and to minimise the development time. In addition, it has to take advantage of every PLC capability, since each PLC has various types of special features, such as Fuzzy Logic controllers, which can be included in an automaton state.

References

[1] Ramadge, P. and W.M. Wonham, (1989) "The Control of Discrete Event Systems", *Proceedings of the IEEE*, Vol.77, No.1, pp.81-98, 1989.

[2] Antsaklis, P. J. and X. D. Koutsoukos, (2003) "Hybrid Systems: Review and Recent Progress", Chapter in Software-Enabled Control: Information Technologies for Dynamical Systems, IEEE Press.

[3] Lee, E. A., and H. Zheng, (2005) 'Operational Semantics of Hybrid Systems', Invited paper in *Hybrid Systems: Computation and Control: 8th International Workshop, HSCC*, LNCS 3414, Zurich, Switzerland, March, 9-11.

[4] Kleijn, J. J. T., Reniers, M. A. and Rooda, J. E. (2003) 'Analysis of an Industrial System', *Formal Methods in System Design*, 22, 249-282.

[5] Lauzon, S. C., Mills, J. K. and Benhabib, B. (1997) 'An Implementation Methodology for the Supervisory Control of Flexible Manufacturing

Workcells', *SME J. of Manufacturing Systems,* Vol. 16, No.1, pp.91-101.

[6] Cassandras, C.G. (1993) *Discrete Event Systems: Modelling and Performance Analysis*, Richard D. Irwin Inc., Boston (MA).

[7] Peterson, J.L. (1981) *Petri Net Theory and the Modelling of Systems*, Prentice-Hall Inc.

[8] Dierks, H. (1997) 'PLC-Automata: A New Class of Implementable Real-Time Automata', ARTS'97, LNCS, Springer Verlag.

[9] Alur, R. and Dill, D.L. (1994) 'A Theory of Timed Automata', *Theoretical Computer Science*, 126:183-235.

[10] Alur, R., Courcoubetis, C., Henzinger, T.A. and Ho, P-H. (1993) 'Hybrid Automata: An Algorithmic Approach to the Specification and Verification of Hybrid Systems', *Hybrid Systems,* Grossman, Nerode and Ravn, Lecture Notes in Computer Science, Vol. 736.

[11] Khoussainov, B. and Nerode, A. (2001) *Automata theory and its applications,* Birkhauser, Boston.

[12] Lygeros, J., Johansson, K. H., Simic, S. N., Zhang, J. and Sastry, S. S. (2003) 'Dynamical Properties of Hybrid Automata', *IEEE Transactions on Automatic Control*, vol. 48, No 1.

[13] Antsaklis, P.J. (2000) 'A Brief Introduction to the Theory and Applications of Hybrid Systems', *Proceedings of the IEEE, Special Issue on Hybrid Systems: Theory and Applications,* Vol. 88, July 2000, pp. 879-887.

[14] Henzinger, T.A. (1996) 'The theory of hybrid automata', In Proceedings of the *11th Annual Symposium on Logic in Computer Science*, pages 278-292. IEEE Computer Society Press.

[15] Balduzzi, F. and Kumar, R. (2003) 'Hybrid Automata Model of Manufacturing Systems and its Optimal Control Subject to Logical Constraints', *International Journal of Hybrid Systems*, pages 61-80, vol. 3, number 1.

[16] Willems, H.X. (1999) 'Compact Timed Automata for PLC Programs', Technical Report CSR9925, University of Nijmegen.

[17] Tapken, J. and Dierks, H. (1998) 'Graphical Development of PLC-Automata', *Proceedings of FTRTFT'98*, vol.1486, LNCS Springer.

[18] Deligiannis, V., S. Manesis and J. Lygeros, (2008). *"Global Automata: a new formal method for modeling industrial systems",* International Journal of Industrial and Systems Engineering.

[19] Lucas, M.R. and Tilbury, D.M. (2004) 'The practice of industrial logic design', *Proceedings of the 2004 American Control Conference*, Boston.

[20] Yevtushenko, N., Villa, T., Brayton, R. K. and Petrenko, A. (2001) ‘Solution of parallel language equations for logic synthesis’, *Proceedings of the International Conference on Computer Aided Design, ICCAD 2001*, San Jose.

[21] Bengtsson, L. and Harju, L. (2002) ‘Modelling and Verification of PLC Control in Buses’, Chalmers University of Technology, Technical Report, Sweden.

[22] Hobfaur, M. and Wotawa, F. (2005) ‘A Diagnosis-Based Casual Analysis Method for Concurrent Hybrid Automata’, *16th International Workshop on Principles of Diagnosis*, Monterey, California, USA, June 1-3.

[23] Zimmerova, B. (2005) ‘Component-Interaction Automata for Component-Based System Specification and Verification’, *LASER'05.*

[24] Pereira, F. C. N. and Riley, M. D. (1997) ‘Speech Recognition by Composition of Weighted Finite Automata’, In *Finite-State Language Processing* (Roche and Schabes, 1997b), chapter 15, pages 431–453.

[25] Deligiannis, V. and S. Manesis, (2005). *“Automata Composition for Modeling Large Industrial Systems”,* CIMCA 2005, International Conference on Computational Intelligence for Modelling Control and Automation.

[26] Pooley, R. (2007). ‘Behavioural equivalence in simulation modelling’, *Simulation Modelling Practice and Theory*, 15 (1): 1-20.

[27] Hendrics, M. and Verhoef, M. (2006), ‘Timed automata based analysis of embedded system architectures’, *20th International Parallel and Distributed Processing Symposium. IPDPS 2006.*

[28] Yalcin, A. and Namballa, R. K. (2005), ‘An object-oriented simulation framework for real-time control of automated flexible manufacturing systems’, *Computers and Industrial Engineering,* 48 (1): 111-127.

[29] Simeu-Abazi, Z. and Bouredji, Z. (2006), ‘Monitoring and predictive maintenance: Modeling and analyse of fault latency’, *Computers in Industry*, 57 (6): 504-515.

[30] Panopoulos, D. and Metaxiotis, K. (2006). ‘Solving production scheduling problems using advanced model checking tools’, *International Journal of Computer Applications in Technology,* 26: 37-48.

[31] Panek, S., Stursberg, O. and Engell, S. (2006). ‘Efficient synthesis of production schedules by optimization of timed automata’, *Control Engineering Practice*, 14 (10): 1183-1197.

[32] Supavatanakul, P., Lunze, J., Puig, V. and Quevedo, J. (2006). 'Diagnosis of timed automata: Theory and application to the DAMADICS actuator benchmark problem', *Control Engineering Practice*, 14 (6): 609-619.

[33] Guo, Y., Liao, W., Cheng, X. and Liu, L. (2006). 'SimOpt: A new simulation optimization system based virtual simulation for manufacturing systems', *Simulation Modelling Practice and Theory*, 14 (5): 577-585.

[34] Rodriguez, S., Hilaire, V. and Koukam, A. (2007). 'Towards a holonic multiple aspect analysis and modeling approach for complex systems: Application to the simulation of industrial plants', *Simulation Modelling Practice and Theory*, 15 (5): 521-543.

[35] Muthiah, K. and Huang S. (2006), 'A review of literature on manufacturing systems productivity measurement and improvement', *International Journal of Industrial and Systems Engineering*, 1 (4): 461-484.

[36] Younis, M. B. and Frey, G. (2006) 'UML-based Approach for the Re-Engineering of PLC Programs', *IEEE IECON 2006 - 32nd Annual Conference on Industrial Electronics,* Paris, pages 3691-3696.

[37] Flordal, H., Fabian, M., Åkesson, K. and Spensieri, D. (2007) 'Automatic model generation and PLC-code implementation for interlocking policies in industrial robot cells', *Control Engineering Practice,* Volume 15, Issue 11, Pages 1416-1426.

[38] Thapa, D., Park, C.M., Dangol, S. and Wang, G.N. (2006) 'III-Phase Verification and Validation of IEC Standard Programmable Logic Controller', *CIMCA Computational Intelligence for Modelling, Control and Automation,* Sydney, Australia.

[39] Åkesson, K., Fabian, M., Flordal, H. and Malik, R. (2006) 'Supremica – An integrated environment for verification, synthesis and simulation of discrete event systems', *Proceedings of 2006 Workshop of Discrete Event Systems (WODES),* Ann Arbor, Michigan.

[40] Music, G., Gradisar, D. and Matko, D. (2005) 'IEC 61131-3 Compliant Control Code Generation from Discrete Event Models', *IEEE International Symposium on, Intelligent Control,* pages 346-351.

[41] Giese, H., Glesner, S., Leitner, J., Schäfer, W. and Wagner, R. (2006) 'Towards Verified Model Transformations', *3rd International Workshop on Model Development, Validation and Verification (MoDeV^{2}a),* Genova, Italy.

[42] Khalgui, M. And Rebeuf, X. (2009) 'An approach to implement a Programmable Logic Controller from real-time software components', *International Journal of Industrial and Systems Engineering,* Vol. 4, No.1, pages 60 – 75.

[43] Sacha, K. (2008) 'Model-based implementation of real-time systems', *Lecture Notes in Computer Science,* Vol. 5219 LNCS, pages 332-345.

[44] Hasdemir, I.T., Kurtulan, S. and Gören, L. (2008) 'An implementation methodology for supervisory control theory', *International Journal of Advanced Manufacturing Technology,* Vol. 36, Issue 3-4, pages 373-385.

[45] Alenljung, T., Sköldstam, M., Lennartson, B. and Åkesson, K. (2007) 'PLC-based implementation of process observation and fault detection for discrete event systems', *Proceedings of the 3rd IEEE International Conference on Automation Science and Engineering,* pages 207-212.

[46] International Electrotechnical Commission (2003) 'International Standard IEC 61131-3 Programmable Controllers – Part 3: Programming Languages', edition 2.0.

[47] Moniruzzaman, M. and Gohari, P. (2007) 'Implementing supervisory control maps with PLC', *Proceedings of the American Control Conference*
2007, Article number 4282159, pages 3594-3599.

[48] Thapa, D., Park, S.C., Park, C.M. and Wang, G.N. (2007) 'Controller logic design and Implementation using t-MPSG', *ICCAS 2007, International Conference on Control, Automation and Systems*, Article number 4406584, pages 1540-1544.

[49] http://www.control.lth.se/~karlerik/grafchart.html.

[50] http://www.uppaal.com.

[51] http://www2-data.informatik.unibw-muenchen.de/DiaGen.

[52] http://www-verimag.imag.fr/TEMPORISE/kronos/.

[53] http://www-cad.eecs.berkeley.edu/~tah/HyTech/.

[54] http://theoretica.informatik.uni-oldenburg.de/~moby/.

[55] http://www.path.berkeley.edu/shift/Default.htm.

[56] Deshpande A., Göllü A. and Semenzato L. "The SHIFT programming language for dynamic networks of hybrid automata*", IEEE Transactions on Automatic Control*, 43(4):584{587, April 1998.

[57] http://www.supremica.org.

[58] Deligiannis, V. and S. Manesis, *"Formal Analysis of Global Automata's Structural Properties",* International Journal of Industrial and Systems Engineering, has been submitted for consideration.

[59] Alur, R., Grosu, R., Hur, Y., Kumar, V. and Lee, I. (2000)'Modular Specification of Hybrid Systems in CHARON', *Proceedings of the 3rd International Workshop on Hybrid Systems: Computation and Control,* Pittsburgh, PA, March 23-25.

[60] Hylands, C., Lee, E. A., Liu, J., Liu, X., Neuendorffer, S., Xiong, Y. and Zheng, H. (2003) 'Ptolemy II - Heterogeneous Concurrent Modelling and Design in Java', vol. 2. Department of Electrical Engineering and Computer Sciences University of California at Berkeley. Memorandum UCB/ERL M03/29, version 3.0.

[61] Deshpande, A., Gollu, A. and Varaiya, P. (1996) 'Shift: a formalism and a programming language for dynamic networks of hybrid automata', *In Hybrid Systems V, Lecture Notes in Computer Science 1567.* Springer-Verlag.

[62] Deshpande, A. and Varaiya, P. (1997)'The Use of SHIFT in System Design', *Conference on Decision and Control,* San Diego.

[63] Antoniotti, M. and Gollu, A. (1997) 'SHIFT and SMART-AHS: A Language for Hybrid Systems Engineering, Modelling, and Simulation', *Conference on Domain Specific Languages*, Santa Barbara, CA, U.S.A.

[64] Deligiannis, V. and S. Manesis, (2006). *"Building Automata-Type Simulation Models for Undergraduate Teaching Industrial Systems Analysis",* ACE '06, 7th IFAC Symposium on Advances in Control Education.

[65] Hamming, R. W. (1973) *Numerical Methods for Scientists and Engineers,* McGraw-Hill, New York.

[66] Kiencke, U., and L. Nielsen, (2000) *Automotive Control Systems for Engine, Driveline and Vehicle*, Springer, London, 2000.

[67] Krutov, V. I. (1987) *Automatic Control of Internal Combustion Engines.* Moscow: Mir Publishers, 1987.

[68] Stone, R. (1992) *Introduction to Internal Combustion Engines.* Hampshire: The Macmillan Press, 1992.

[69] Taylor, C. F. (1985) *The Internal-Combustion Engine in Theory and Practice.* Massachusetts: The M.I.T. Press, 1985.

[70] Heywood, J. B. (1988) *Internal Combustion Engine Fundamentals.* New York: McGraw-Hill, 1988.

[71] Kuo, P. S. (1996) "Cylinder Pressure in a Spark-Ignition Engine: A Computational Model", Engineering Sciences, 1996.

[72] Ferraris, C. (2001) "Concrete Mixing Methods and Concrete Mixers: State of the Art", *Journal of Research of the National Institute of Standards and Technology,* NIST, vol. 106, pp. 391-399, 2001.

[73] Setien, J., I. Carrasal, J. Figueroa and J. Polanco, (2003) "Application of an artificial neural network to ready-mixed concretes mix design," *Journal of Materiales de Construccions*, Castineira Libreria Tecnica, vol. 53, issue 270, pp. 5-19, 2003.

[74] Zayed, T. and D. Halpin, (2001) "Simulation of concrete batch plant production," *Journal of Construction Engineering and Management,* ASCE, vol. 127, pp. 132-141, 2001.

[75] Sawhney, A., O. Abudayyeh and T. Chaitavatputtiporn, (1999) "Modeling and Analysis of Concrete Production Plant Using Petri Nets", *Journal of Computing in Civil Engineering,* ASCE, vol. 13, No. 3, pp. 178-186, 1999.

INDEX

A

B

C

D

E

F

G

H

I

J

K

L

M

N

O

P

R

S

T

U

V

W

X

Y